Packaging for the Environment

A Partnership for Progress

"We can establish a new kind of environmentalism, one in which a sound ecology and a strong economy go hand in hand."
—George Bush, President of the United States

"What is needed now is a new era of economic growth—growth that is forceful and at the same time socially and environmentally sustainable—it is possible to join forces, to identify common goals, and to agree on common action."
—Gro Harlem Brundtland, Prime Minister of Norway

"Corporations that think they can drag their heels indefinitely on genuine environmental problems should be advised: society won't tolerate it, and Du Pont and other companies with real sensitivity and environmental commitment will be there to supply your customers after you're gone."
—Edgar S. Woolard, Chairman of Du Pont

"Our old environmental foe, modern technology, must become a friend . . . economic growth has its imperatives; it will occur. The key question is: with what technologies?"
—James G. Speth, President of the World Resources Institute

"Furthering technological and economic development in a socially and environmentally responsible manner is not only feasible, it is the great challenge we face as engineers, as engineering institutions, and as a society."
—Paul E. Gray, President of the Massachusetts Institute of Technology

Source references for these quotes are given at the back of the book.

Packaging for the Environment

A Partnership for Progress

E. Joseph Stilwell
R. Claire Canty
Peter W. Kopf
Anthony M. Montrone

Arthur D. Little, Inc.

American Management Association

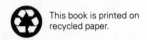
This book is printed on
recycled paper.

This book is available at a special discount when ordered in bulk quantities. For information, contact Special Sales Department, AMACOM, a division of American Management Association, 135 West 50th Street, New York, NY 10020.

This publication is designed to provide accurate and authoritative information in regard to the subject matter covered. It is sold with the understanding that the publisher is not engaged in rendering legal, accounting, or other professional service. If legal advice or other expert assistance is required, the services of a competent professional person should be sought.

Library of Congress Cataloging-in-Publication Data

Packaging for the environment: a partnership for progress/E. Joseph Stilwell . . . [et al.].
p.   cm.
Includes bibliographical references and index.
ISBN 0-8144-5074-1 (hardcover)
1. Package goods industry—Environmental aspects. 2. Package goods industry—Environmental aspects—Case studies. 3. Recycling (Waste, etc.) 4. Recycling (Waste, etc.)—Case studies.
I. Stilwell, E. Joseph.
TD195.P26P33 1991
688.8—dc20   91-55507 CIP

Contents

*All case studies are published with the permission of the companies involved.

Preface

The packaging industry has been on the "frontline" in the municipal solid waste issue for several years. Corporations representing both the supply and demand side of this industry have formed and tested many strategies; the key stakeholders have been identified and their positions well defined; and the issues have been structured, debated, and restructured many times. There is a great deal of learning that is taking place in this industry that would be of benefit to all industries producing products and creating waste.

Through our work with these issues, we have come to realize that while the debate flourishes, there is a great deal that is not being said. A single issue mentality prevails and there are only isolated attempts at consensus building among key stakeholders. Consequently, we decided to help bridge the communication gap by writing a book that would address the issues in all their complexity and illustrate that through partnering and consensus building, there are achievable solutions that are both sound business and can have positive impact on the environment.

We hope that this book will dispel some of the common environmental myths surrounding packaging and its component materials, and show how sound packaging principles fit within a larger program of corporate environmental responsibility. While the issues are challenging and complex, so too is the understanding necessary to communicate these issues. Only with the unselfish assistance of several already too busy Arthur D. Little professionals could we have hoped to achieve our goals. We would like to acknowledge the special skills and expertise of those who have made this work possible:

Andrea Durham, a consultant and packaging engineer, contributed packaging materials and industry insight from her experience in both consumer and industrial packaging.

Tom Hambleton, a senior consultant in packaging, contributed his special knowledge of package development and his extensive industry experience to many sections of the book.

Mike Rossetti, a ceramic engineer and a senior consultant in Materials and Applied Physics, framed our comments on glass as a packaging material and as a supply side industry.

Viktors Vejins drew from his extensive technical knowledge and management consulting experience in metals and materials industries to form major portions of the materials and infrastructure sections of the book.

Dr. James Racich, a senior consultant in plastics and composites, contributed greatly to the materials section of the book and assisted with industry interviews.

Dr. Stephen Rudolph, Vice President and Managing Director of Product Technology Laboratories, drew from his 20 years' experience in organic and polymer chemistry to enrich several sections of the book.

Nancy Smith, Vice President and Managing Director of Food Industries, contributed her insight in food formulation, preservation, and marketing.

Dr. Charles Kusik, a director in Environmental, Health, and Safety, drew from his background in chemical engineering and his experience in environmental issues to help form the infrastructure portion of the book.

Dr. Terry Rothermel, a senior consultant in Waste Management, with 25 years' experience in management and market consulting led the development of the infrastructure section of the book.

Ralph Earle, senior consultant in Waste Management, drew from his rich experience in public sector policy development to help describe the current state of the issues with emphasis on integrated waste management and market-based environmental solutions.

Karen Blumenfeld, a senior consultant in Waste Management, drew from her 15 years' experience in both the public and private sectors to help us frame our description of the issues, the waste management infrastructure, and the tools for industry solutions.

Alan Murphy, Director of Cambridge Consultants, Limited (an Arthur D. Little subsidiary headquartered in Cambridge, England), provided the summary of Europe's environmental history and described current issues and solutions.

Hiroyuki Sakemi, associate consultant in Arthur D. Little's Tokyo office, provided an update on the environmental experience in Japan.

Hardin Tibbs, an industrial designer by training and a packaging and graphics communication designer by practice, led the framing of our thinking on the emerging industrial ecology, assisted in conducting industry interviews, and managed the production of the book—he also is responsible for the cover design.

This book would be little more than a collection of observations and theories if we had not been joined by five major multinational corporations who have been pioneering proactive environmental leadership. We wish to thank each of these corporations for allowing us to print their stories with special thanks to the individuals within these corporations who very candidly discussed their policies, practices, and the impact that environmental strategies have had upon their corporate culture, their operations, and their markets:

3M Company: Dr. Robert P. Bringer, Dr. Daniel J. Knuth, Dr. Donald R. Theissen, Thomas W. Zosel, and Richard H. Renner.

E. I. du Pont de Nemours & Company: Frank Aronhalt and Richard Steward.

Dow Chemical Company: Ken Harman, Angee Linsey; and at the National Polystyrene Recycling Company, Ellen Kousty.

Johnson & Johnson: Dr. Peter N. Britten and Dr. Lawrence G. Mondschein.

We would like to thank the several people at Procter & Gamble Company who assisted in development of their story.

And a special thanks from all who shared in this project to Huguette Shepard, who kept communications intact and events on schedule, and who was our first line of defense against errors and oversights.

The authors wish to dedicate this work to Catherine Marenghi, who through her considerable skill and patient understanding, created a readable format and style from our rather stilted and often unnecessarily complex jargon. We would also like to thank her son Steven who demonstrated equal patience and understanding by waiting until the draft was finished before arriving in this world.

Introduction

Packaging is the ultimate symbol of our consumer culture. It tells the story of our technological achievements, preserves our food, protects what we buy, and raises our standard of living. It plays a vital and growing role in the global economy. And through Andy Warhol's vision, the Campbell's soup can and the Brillo box have even been elevated to an art form. At the same time, packaging is also the largest single contributor to one of the nation's most troubling environmental problems: the municipal solid waste crisis.

Simply stated, we are running out of room for our garbage. Landfills all over the world are literally spilling over; in the United States alone, one-quarter of the country's municipalities are expected to exhaust their landfill capacity before 1995. And the oceans, once treated cavalierly as an infinite waste depository, have begun to wash back our own refuse to our shores. The 1990s will be a decade of reckoning for solving the solid waste crisis.

As disposal capacity shrinks, uneasiness about the crisis is escalating dramatically. Surveys have indicated that the American public ranks environmental concerns among the most important problems facing the country—more important than crime, unemployment, AIDS, and other prominent issues. Just a few years ago, the environment was rated as low as fourteenth among the nation's most pressing problems. [1]

Packaging is the largest, fastest growing, and most complex component of municipal solid waste (that is, municipal waste generated by homes and businesses, as opposed to industrial and agricultural waste). Between 1958 and 1976, per capita consumption of bottles, cans, boxes, wrappers, and other packaging increased by 63 percent in the United States. In 1990, packaging accounted for more than 30 percent of the municipal solid waste stream (see Figure I-1).

Packaging thus creates a unique opportunity for industry to assume a leadership role in solving the municipal solid waste crisis. Virtually every industry has a stake in packaging—whether as manufacturer, distributor, user, or seller. And because of the sheer volume of packaging in the municipal solid waste stream, even incremental improvements in packaging can make a real difference in solving the garbage crisis.

Businesses can and must be responsible practitioners of environmentally sound packaging. The reasons for assuming such a role are not merely altruistic. A great many major corporations are building market share and achieving a competitive edge by demonstrating environmental consciousness in their packaging policies.

Here are some steps companies have taken in recent years:

Figure I-1. Components of municipal solid waste and types of packaging waste, by percent.

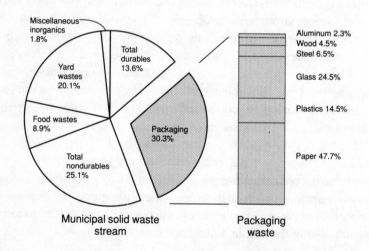

Miscellaneous inorganics 1.8%
Yard wastes 20.1%
Total durables 13.6%
Food wastes 8.9%
Packaging 30.3%
Total nondurables 25.1%

Municipal solid waste stream

Aluminum 2.3%
Wood 4.5%
Steel 6.5%
Glass 24.5%
Plastics 14.5%
Paper 47.7%

Packaging waste

- Coca-Cola Company and PepsiCo, Inc., announced simultaneously that they would introduce bottles made with reused plastic—becoming the first beverage companies to reuse plastics for containers.

- Procter & Gamble has pioneered a number of environmental improvements in packaging, including the first major consumer product in the United States—Spic & Span—in a 100 percent recycled plastic bottle. P&G has reduced the amount of packaging it produces, by offering refills and concentrates, and it is committed to using at least 25 percent recycled content in its plastic bottles. Most important, it has helped develop an infrastructure for recycling (see case study, Chapter 9).

- H. J. Heinz Company replaced its multilayer plastic ketchup bottle with a polyethylene terephthalate (PET) bottle that is more easily recyclable. The company expects to improve its 51 percent share of the $600 million American ketchup market by prominently labeling the bottle as recyclable.

- The Body Shop, a British-based retailer of natural and organic hair- and skin-care products, has differentiated itself through "green marketing." Its stores display information on environmental issues like ozone depletion and global warming, provide refillable containers, announce that no animals are used in product testing, and offer discounts to consumers who reuse the company's packaging. Even its catalogs are recycled; mail orders are packed in shredded Body Shop catalog pages.

- International Paper has invented a paper package that not only is 10 percent cheaper than plastic containers but also extends the shelf life of juices, soups, cereals, and other products. The composite paper package is designed to be recyclable.

The idea that the consumer will reward companies for environmental consciousness was aggressively stated by Edgar S. Woolard, the chairman of Du Pont:

industrial companies will ignore the environment only at their peril . . . . Corporations that think they can drag their heels indefinitely on genuine environmental problems should be advised: society won't tolerate it, and Du Pont and other companies with real sensitivity and environmental commitment will be there to supply your customers after you're gone. [2]

Scope of the Problem

How serious is the municipal solid waste crisis? The following are a few indicators:

• According to the United States Environmental Protection Agency, municipal solid waste generation in the United States has doubled in less than three decades, from 88 million tons in 1960 to nearly 180 million tons in 1988—4 pounds of waste per person per day. And the garbage is mounting at a much faster rate than the EPA had earlier predicted. As recently as 1986, the agency forecast that waste generation would not reach 4 pounds per person per day until the year 2000; the EPA now projects that by 2000 the United States will generate 216 million tons per year, almost 4.5 pounds per person per day (see Figure I-2).

• American consumers are dumping garbage at a much faster rate than they are recycling, but the ratio is improving. In 1988, according to the EPA, 72.7 percent of the nation's municipal solid waste ended up in landfills and 14.2 percent was incinerated. In 1988, 13.1 percent of this country's waste was recycled, compared with 7.2 percent in 1960.

Figure I-2. Municipal solid waste generation in the United States, historical and EPA projection.

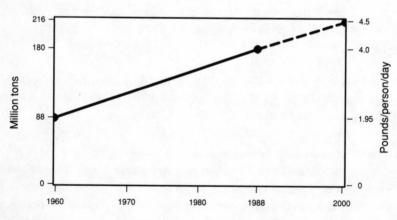

Source: U.S. EPA

• Serious efforts at recycling have only just begun. Together, Waste Management, Inc., and Browning Ferris Industries supply curbside recycling to more than 50 percent of the American homes that have such services. But only 9 million homes across the nation have recycling services, and 80 percent of these homes are concentrated in the Northeast. While half of the aluminum cans sold in the United States are recycled, only 8.5 percent of all glass containers are recycled, and only 1–2 percent of all plastic containers.

• More than half the population of the United States lives in regions with fewer than ten years of landfill capacity left. By 1995, all landfills in New York State will have reached their capacity and will have to close down, according to the New York State Legislative Commission on Solid Waste Management. [3]

• The average "tipping fee" that waste haulers pay to unload at disposal sites increased 250 percent, from $10.80 per ton in 1982 to $26.93 per ton in 1989.

The Message for Business

Of course, industry is not alone in creating the municipal solid waste crisis; the responsibility is shared by consumers, government, and industry alike. In fact, some business leaders might argue that they are only responding to consumer demand when, for example, they deliver microwave dinners in single-serving disposable packages. However, businesses can no longer afford to be merely reactive to market pressures, government regulations, or activists' protests. The challenge for businesses is to balance the public's desire for convenience, attractiveness, price, and quality with environmentally sound practices. At stake is their credibility, their competitiveness, and, ultimately, the "bottom line."

The trend toward environmentally safe products and packaging is not a fad. Sometimes publicly, sometimes quietly, many corporations are taking a proactive role in defining the outlines of an emerging "industrial ecology." This is a philosophy which holds that environmental consciousness and sound business practices are not mutually exclusive. Businesses can and do make money on environmentally responsible packaging: by avoiding future liabilities, by building credibility with customers, and by tapping new business opportunities.

The "business as usual" approach will no longer be tolerated. Public alarm over the garbage crisis is sending a clear message to businesses: "Clean up or else."

The Role of Packaging

Truly "bad" packaging from an environmental standpoint is a matter of extremes: underpackaging is disastrous because it does not preserve and protect; and overpackaging is equally problematic for its wastefulness and cost.

Packaging is the tool that protects and contains goods so that the environmental impact of our consumption is minimized. It is vital to the health and welfare of the consumer. However, packaging also consumes resources, and raw materials and energy are costly to manufacturers. It is therefore in the self-interest of the manufacturer to reduce waste at the source. "Lightweighting" is a long-standing trend in packaging; originally adopted in order to reduce energy and materials costs, it has additional environmental benefits as well.

Most households use dozens of packaged items every day, and two-thirds of all packaging is used to protect food. In developing countries, between 30 and 50 percent of food shipments are spoiled because of inadequate storage and distribution systems. In more developed countries with sophisticated packaging and distribution, only 2–3 percent of food shipments are wasted.

The growth of packaging has paralleled a change in the way products are sold and marketed. Because of rising labor costs, the retail industry moved in the direction of discount and self-service department stores. As a result, product packaging came to assume a new role: Informative package labels with pictures took the place of the sales clerk. In many cases, this new marketing function meant a different form of packaging: Consumer goods were no longer manually wrapped in tissue and boxed by a salesclerk but were packaged at the factory in display boxes that went from retail shelf to home. This was a major revolution in product packaging (see Figure I-3).

More recently there have been other revolutionary developments in packaging. Product tampering, as seen in the Tylenol cyanide poisoning scare of 1982, brought a new wave of packaging innovation to develop tamper-evident containers. And we are now seeing the latest revolution in packaging: both through materials and design, packaging that will have the gentlest impact on the natural environment.

While the sheer volume of municipal solid waste is the most obvious and visible problem for the environment, the public is rightly concerned about many related environmental problems: the toxicity of landfills, the depletion of scarce or nonrenewable resources, litter, pollution, ozone depletion, and wildlife endangerment. Let us briefly examine the history of these concerns, as well as the industry and legislative responses.

Growing Concerns

Environmentalism is a comparatively recent phenomenon. In its earliest incarnation, it was known as conservation. President Theodore Roosevelt, an avid hunter, was an unlikely but ardent proponent of wildlife conservation in the early 1900s.

Figure I-3. The evolution of trends in the packaging industry.

The '60s	• Convenience as a basis of competition—"throw away society" • POP marketing—"the silent salesman"
The '70s	• Fragility assessment—analytical design for movement • Energy shortage—lightweighting (the first source reduction)
The '80s	• Malicious tampering—"safe" packaging • Extended shelf life—barrier technology; controlled atmosphere technology • Quality—sensory issues • Industry restructuring—"evaporation" of resources
The '90s	• Environmental issues

Concern for the environment on a broader scale took longer to emerge, and environmentalism was for a time widely seen as a left-wing cause. But it is now supported by the full spectrum of political movements, age groups, and other constituencies. The "throw-away society" that began in the 1950s and 1960s, the energy crises of the 1970s, and the groundswell of environmental concern that formed in the 1980s has set the stage for what may now be termed the "green 1990s."

Pesticides

The first wave of intense environmental concern in the United States came with the 1962 publication of Rachel Carson's *Silent Spring*. That book raised concern about the environmental damage caused by the use of DDT and other pesticides. Perhaps the most poignant effect of DDT was its damage to the reproductive system of the bald eagle, our national bird. Throughout the 1960s and 1970s, pesticides and their threat to both humans and wildlife were a major target of the government, the media, and environmentalists.

Anyone who doubts the relevance of environmental issues to business need only study the case history of Northwest Industries and the company it acquired in 1964, Velsicol Chemical Corporation, a maker of agricultural chemicals. Because of growing awareness of the environmental impact of pesticides, the parent company faced liabilities nearing one-third the book value of Velsicol. Harder to measure was the long-term damage to the company's credibility. As Ben E. Heineman, chairman of Northwest Industries, put it, "Had we anticipated the [environmental problems] correctly, we probably would not have bought Velsicol." [4]

Litter

Another great public campaign of this period was waged against litter. "EVERY LITTER BIT HURTS" was the rallying cry of what was positioned largely as an aesthetic issue: Litter was ugly. Litter was also expensive to clean up and dangerous to wildlife. Beverage containers were one of the products most directly affected by this initiative, and the result was the imposition of mandatory deposit laws in eleven states.

The fast food industry was a major target of the litter cleanup campaigns, and such restaurants were banned from a number of communities that were concerned specifically about litter. The fast food business was singled out in the 1970s as an extravagant waster of paper, used in burger wrappers and French fried potato boxes, cups, bags, napkins, take-out trays, and even luncheon tray liners. During the 1970s, a decade of serious paper shortages, restaurant chains like McDonald's became the object of consumer boycotts and "Save a Tree" campaigns.

In response to environmental concerns, McDonald's established its own voluntary recycling network for food containers and funded a major public education initiative. Interestingly, the early public pressure over the issue of paper waste led McDonald's to switch to a polystyrene foam sandwich container, the so-called clamshell—which was later the object of more vehement consumer protests. The company waged an uphill battle to assure the public of the environmental safety and recyclability of the clamshell, but finally reacted to consumer pressure by switching to a composite paper wrap—which, ironically, will be difficult to recycle. It was a business decision based not on technical merits but on public perceptions. (See case study on National Polystyrene Recycling Company, Chapter 11.)

Other Concerns

Ozone depletion has been linked to a number of packaging materials and products. Polystyrene foam was once produced by a process that used chlorofluorocarbons (CFCs) as blowing agents; the process released CFCs into the atmosphere, where their photochemical byproducts erode the earth's protective ozone layer. CFCs used as refrigerants and aerosol spray propellants are also linked to ozone damage. In response to this serious problem, industry has implemented alternative manufacturing processes, developed alternate blowing agents, or switched to other materials. Aerosol spray containers use alternative propellants, and have been widely replaced by pump-spray bottles and other forms of packaging. And the recycling of polystyrene foam is emerging as a young and promising industry. Nevertheless, polystyrene products continue to face public resistance.

Wildlife endangerment is another major issue with a history of direct impacts on packaging. For example, it was learned that pull tabs on beverage cans, when discarded in waterways, attract fish with their bright metallic shine; if swallowed, they can cause death or injury because of their sharp edges. Land animals also face the danger of swallowing the sharp tabs. As a result, pull tabs were widely banned, replaced by "ecology lids" with no throw-away parts. Plastic carriers for six-pack cans have also been identified as a threat to animals, especially waterfowl, that might be choked or strangled by them. These have been widely replaced by degradable carriers, and state bans are in effect against the nondegradable variety.

Resource conservation is an environmental concern that has risen and fallen over the years, often in proportion to the severity of successive energy crises. Conservation is an issue that cuts across all types of packaging. The industry response has been to minimize product packaging and to voluntarily recycle packaging such as aluminum beverage cans. On the legislative side, the 1970 Resource Recovery Act resulted in extensive research and pilot projects. The 1976 Resource Conservation and Recovery Act addressed source reduction through product regulation and financial incentives and disincentives. This act recommended further study, but these recommendations were not acted on and no specific action followed.

Finally, the municipal solid waste crisis is emerging as a top environmental concern. The risk is perceived on two fronts: landfill capacity shortage and exposure to toxins, which can leak from landfills into water supplies. All types of packaging contribute to the municipal solid waste crisis, but especially those that are highly visible, have a short lifespan, are produced in high volume, are growing at a fast rate, are excessive or unnecessary, and are potentially toxic.

Responding to the Municipal Solid Waste Crisis

The industry response to the municipal solid waste crisis has been multifold and includes the following initiatives:

- Research and development.
- Joint ventures with recyclers.
- Public education/promotion.
- Voluntary labeling.
- New products and packaging concepts.

Legislative initiatives have paralleled the industry's work in public education, R&D, and product labeling and have additionally provided a system of financial incentives and disincentives to both industry and consumers. Not only has the public sector supported curbside recycling, but in many cases it has imposed mandatory source separation laws and even fines for noncompliance. In New York City, uniformed "recycling cops" from the city's sanitation department patrol the neighborhoods on foot, sifting through the residents' garbage and writing up tickets for such infractions as discarding clean newsprint. Fines start at $25, and persistent violators can be fined as much as $10,000.

Subsequent chapters will provide more detailed case histories of industry's efforts to manage environmental concerns.

The Seven Key Players

Seven key groups are providing the driving forces behind the move toward "environmentally friendly" products and packaging. They are environmental activists, the media, the general public and consumers, educators, government, retailers, and finally the packaging industry itself.

We should begin by looking at the packaging industry. Not an industry in the typical sense, it defies strict definition. It includes product marketers, product and packaging manufacturers, and raw material suppliers and involves a wide variety of materials, technologies, establishments, and institutions.

The relationship of the packaging industry with the other key drivers of environmentally sound packaging is often adversarial and has historically been reactive in nature, although the industry is not alone in its reactive posture. The public reacts to the media. Legislators and the media react to environmentalists, often without any rigorous technical proof or evidence. Environmentalists react to industry. Industry reacts to perceived market pressures, which it often cannot verify, and to the regulatory process of incentives and disincentives. Regulators themselves react to many of the same pressures from the public and activists by imposing penalties and restrictions, but offer little in the way of positive guidance. Even educators have played the role of the adversary, encouraging their students to voice protests against corporations and public officials.

The general public believes itself to be environmentally conscious. Surveys show that up to 78 percent are willing to change their purchasing behavior because of environmental concerns. Depending on the surveys, the numbers vary widely. Between 54 and 80 percent report that they already have changed their behavior, and up to 77 percent say that their purchases have been influenced by companies' environmental reputations.

Nevertheless, basic product attributes such as quality, pricing, and convenience still dominate as the principal determinants of consumer purchasing behavior. Industry will be well-advised not to rely too heavily on consumer attitude surveys. Professor William Rathje's famous "Garbage Project" has shown, through his study of waste streams from selected communities, that there is a gulf between word and deed, between what consumers say and what they do.

Consumers are most likely to be influenced by the media and by national environmentalist groups such as Greenpeace, the Sierra Club, and the Audubon Society, as well as by local grass-roots efforts. By working more closely with the media and environmentalists, industry could, in turn, exert a more positive influence on the consumer. Unfortunately, an adversarial relationship may exist between environmentalists and corporations. And few companies have the public relations skills to educate the media in a sophisticated way, and many executives feel they have been burned by the press.

For industry to influence consumer behavior in a meaningful way, it must play a more direct role in educating the consumer. For example, even though Heinz introduced a more environmentally friendly ketchup bottle, its development efforts could not yield environmentally positive results if those PET bottles ended up in landfills. Heinz therefore decided to accompany its new package with a public awareness campaign designed to educate the consumer about PET products and the need to recycle them.

Most of the key players in the waste management arena seek to influence legislation. In many ways, Europe is far ahead of the United States in terms of environmental political movements "green" parties, and American industry can learn much from observing the European scene. Within the United States, the Coalition of Northeastern Governors, among other organizations, has provided leadership on a number of environmental initiatives, working closely with environmental groups and industry.

Among the seven key players there exists a loose partnership of sorts. Sometimes together, more often separately, these groups are working toward defining concepts and standards for environmentally sound packaging. For years, the packaging industry has sought to outrun or defeat regulations promulgated after others have influenced the political process. It can no longer afford to do so. Industry must build upon and strengthen its relationships with the key players in the process, and work within this loose partnership to define a new industrial ecology.

"Going environmental" need not have negative business impacts. Indeed, for the packaging industry there is no longer any other choice.

Strategy for Corporations

It can be argued that the public's perception of an environmental hazard can be more powerful than the hazard itself. Certain packaging practices and materials may come into vogue or fall into disfavor without any basis in logic or in technical evidence.

For example, what does a corporation do when the public perceives that "all plastics are bad," despite substantial evidence to the contrary? Should the company respond to consumer pressures and abandon plastic packaging? Or should it go against the tide and adopt a program of research and education to set the record straight?

This is the strongest reason possible for industry to adopt a proactive—or better, a preemptive—role in defining an "industrial ecology." Industry should define the terms of the environmental debate, rather than take the more difficult route of attempting to change opinions that are already strongly entrenched.

The problem is complicated by the fact that environmentalism is a young field, and we are still in the process of discovering the long-term environmental consequences of our products and packaging. Today, our information may tell us that plastic is the best packaging material; tomorrow, a new discovery may point in the direction of paper packaging. How does industry "do the right thing" when the rules of the game are constantly shifting?

Is paper better than plastic? The methodologies of life cycle analysis, used to evaluate the environmental profiles of products and packaging, are still being refined. At one time, paper packaging was thought to be the most environmentally sound packaging material, because it is biodegradable. But paper does not biodegrade in modern landfills, which seal out the moisture and air that are needed to break down paper. Hence, plastic has come to be regarded as more recyclable and reusable, as taking less energy to produce and doing more to conserve natural resources. And because they are derived from oil and natural gas, some plastics have energy content near that of fuel oil. As we learn to control incineration and manage toxic emissions, siting problems for incinerators may change, and plastics may be in demand as energy sources.

In ten years, if compost siting is not a political or practical problem, paper may then be seen as the most sound material. And we will again need to revise our assumptions about what is environmentally sound.

The answer to the paper versus plastic question is that there are no easy answers. However, industry can and must make a commitment to environmentalism. In this book we will prescribe a strategy for practicing an industrial ecology designed to have the gentlest effect on the natural ecology, where the two interface, while helping companies stay profitable. We will show, through case studies, how decision makers have made situational decisions within a long-term strategic framework.

Again, the detailed strategy of today may not be the same as the strategy of tomorrow. That is why industry is well-positioned to assume a leadership role. American industry has a long history of adapting to consumer tastes and needs. It is in the business of forming and implementing strategies in a fast-changing marketplace—before scientists, legislators, or anyone else can. In this book we will offer an approach for turning negatives into positives—steps that are identifiable and *achievable*—to adapt to change.

In the following chapter we will explore in greater detail the roles of the key stakeholders in the complex process of finding environmental solutions. Later chapters will describe in detail the packaging industry, the materials and technologies that go into packaging, the infrastructure of waste management, and insights from the international scene. Case studies will demonstrate strategies and methodologies that any enterprise can adopt to implement an industrial ecology and establish the sort of partnership that is essential to progress.

Notes

1. Cambridge Reports and Arthur D. Little.
2. Remarks made at the World Resources Institute, Washington, D.C., December 12, 1989.
3. National Solid Waste Management Association, "Public Attitudes Toward Garbage Disposal," 1987.
4. John Diebold, *Making the Future Work* (New York: Simon & Schuster, 1984), p. 237.

1

Key Players in the Environmental Arena

The packaging industry faces a rapidly changing and uncertain business climate in the United States. Multiple constituencies are creating pressure for environmentally responsible products and packaging. Seven main groups influence the pressures for "environmentally friendly" products and packages. The key stakeholders involved are environmental activists, the media, the general public or consumers, educators, government at all levels, retailers, and, finally, the packaging industry itself.

Each of these groups brings its own unique history, its own concerns, and its own agenda to the table. Each is working, in its own way, toward defining concepts and standards for environmentally sound packaging. Understanding their diverse perspectives will be a prerequisite to achieving a working partnership for the future (see Figure 1-1).

Consumers

Citizens of every age group, every political party, and every region have been moved by the sight of oil spills contaminating our shores and by alarming reports of a growing "hole" in the atmosphere's ozone layer that

Figure 1-1. The key stakeholders.

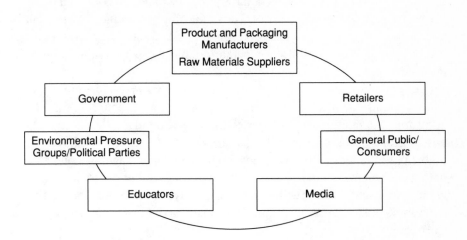

protects us from damaging ultraviolet radiation. In 1987, the infamous trash barge that left New York's Long Island with no place to unload its cargo, turned away from one port after another, became a highly visible metaphor for an environmental crisis that would not go away.

A segment of the American public has responded by embracing recycling, separating trash, supporting environmental activism, and favoring "green" products. Today, the American public believes itself to be environmentally conscious, and most consumers say they are willing to change their purchasing behavior to help improve the environment. A survey by the Michael Peters Group found that 77 percent of those questioned would willingly pay a higher price for a product packaged with recyclable or biodegradable materials. Another study, by the Food Marketing Institute, found that about two-thirds of those surveyed would be influenced by the fact that a product's packaging was made from recyclable materials.

However, while environmental issues are important to the American public, there is a limit to what consumers will do to change their behavior. The consumer is equally or more likely to be concerned with the price and safety of a product than with its packaging, and product convenience is a critical consideration. A survey by Opatow Associates found that currently consumers do not believe take-out food containers are enough of an environmental problem to give up the convenience of leak-proof, heat-retaining packages. And recycling efforts falter when convenience becomes an issue. Curbside recycling is preferred over convenient drop-off centers, 66 percent to 26 percent, according to the Food Marketing Institute. Moreover, curbside recycling programs often generate double the participation rate of drop-off programs.

Part of the problem is that the public does not always link the municipal solid waste problem with their purchasing decisions. Accounting for 30.3 percent of the municipal solid waste stream, packaging is the largest and fastest-growing contributor to the garbage crisis. For the present, however, municipal solid waste disposal is low on the consumer's agenda compared with the problems of nuclear waste disposal, water pollution, toxic wastes, air pollution, and the destruction of the ozone layer, which were the top five environmental concerns according to a recent survey by Opinion Research Corporation.

In spite of good intentions, there is a great deal of public confusion over what is environmentally sound. For example, degradability has been perceived as an attractive solution. The environmentally conscious supermarket shopper, when given the choice of paper or plastic grocery bags, will often choose paper, confident in the belief that paper will biodegrade in landfills. Relatively few consumers know that virtually nothing degrades in today's airtight, watertight landfills. Both types of bags are, in fact, recyclable, and supermarkets may try to promote recycling by providing collection bins for returned bags of either type. But it is difficult to shake the consumer's long-cherished beliefs in degradability.

Because of the complexity of environmental issues and the competing and sometimes misleading claims of "green" marketers, consumers are likely to become confused and to seize upon single issues and simple concepts. Ideas such as "degradability is good" or "plastics are bad" become firmly entrenched. Consumers do not have the time or the resources to look at the full life cycle of a product and to study all of its environmental implications. In the absence of easy-to-follow guidelines based on sound environmental thinking, the best-intentioned consumers are likely to make the wrong choices. For example, the Food Marketing Institute found that consumers continue to prefer conventional laundry detergent packaging to waste-minimizing concentrates and refills in reusable packaging. It is to be hoped that such consumer packaging preferences will change as the environmental consequences of purchase decisions are better understood.

Educators

There is a tremendous amount of environmental education activity at the elementary school level, especially in grades 3–5. Children are a receptive audience to environmental messages on television and in the classroom and are enthusiastic participants in recycling programs. At the high school level, science teachers often deliver environmental themes both in course material and through increasingly popular extracurricular environmental clubs.

Nearly every college and university offers courses relating to the environment, and some offer degrees in environmental science. Urban studies programs of the 1970s have been redesigned with an environmental focus. As state colleges have reduced their emphasis on teacher preparation, environmental programs have been getting more attention. And traditional degree programs within the liberal arts have taken on new disciplines; for example, there are programs in environmental economics, environmental statistics, and environmental law. Packaging design programs now incorporate environmental considerations into course offerings.

Both directly and through trade associations, industry has provided a variety of tools and grants to educators, often aimed at the grade school level. Du Pont provides one example of an effective program of "educating the educators"—not by designing programs itself, but by providing resources for teachers to do so (see Chapter 8).

The Media

Consumer behavior is highly influenced by the print and electronic media, particularly television. The "fourth estate" has a clear and vital role to play in creating a society that is informed and responsible on environmental issues.

But how well-equipped are the media to assume this powerful role? Like the consumer, the typical reporter is barraged with information from multiple sources, all competing for his or her attention. A single reporter may receive several hundred press releases in a day, not to mention the steady stream of phone calls and letters from publicity agents. From these competing information sources, the reporter must sift out the most important stories, always under the pressure of a deadline.

The television journalist faces the additional pressure of competing for ratings, and what "sells" is not always the most significant story. The story of a whale trapped by an arctic ice floe may run for several nights on the evening news, but a major chemical corporation's decision to halt chlorofluorocarbon (CFC) production will not make the TV news at all. Which story has the greater impact on the earth's environment? As long as the business of TV journalism is drama, the highly visual human-interest story will win out over the complex, conceptual story.

The media derive their environmental information both from activist groups and from industry sources, but the activists have been particularly effective in this role. The packaging industry to date has had an uneven record in communicating important issues to the press. The paper industry has one of the most positive environmental stories to tell in its reforestation work, particularly with respect to the Southern pine. And yet, these quiet successes go unreported. The result is the unsettling public perception that industry is the enemy of forest conservation. By failing to play an effectively proactive role with the press, industry is forced into assuming a defensive posture.

Eco-activism is fashionable in the entertainment community. Companies like Du Pont are providing technical expertise to Hollywood consultancies that directly influence how environmental concerns are portrayed in television and film productions. Ideally, consumers will respond to what they learn from the popular media. As more attention is paid to the problem of municipal solid waste by an informed media, more consumers are better equipped to make informed choices about the environmental implications of products and packaging.

Earth Day 1990, marking the twentieth anniversary of the first Earth Day observance, was a powerful catalyst in drawing media attention to the environment. During 1990 a ten-part public television series entitled *Race to Save the Planet* examined every facet of environmental concern on a global scale. Around the same time, new publications sprang up, and established publications devoted advertising and editorial inserts to environmental issues. Hundreds of books were published, many of them how-to books for consumers, all hoping to ride the "green" wave.

That momentum from Earth Day has continued. Environmentalism is garnering increasing coverage by all the media. The environmental reporter is a new and welcome phenomenon, and environmental publications in the United States alone now number several hundred. As this media specialty matures, and as the packaging industry becomes more sophisticated in communicating with the press, the consumer can only benefit.

Environmental Groups

While once perceived as a radical fringe concept, environmentalism has won broad popular support across the country. As a result, the focus of environmental activist groups has changed considerably over the years. From the time of the Industrial Revolution, conservation of natural resources has been a recurring theme, supported by such diverse voices as hunters and bird watchers, botanists and farmers. Teddy Roosevelt, the Audubon Society, the Sierra Club, and Greenpeace are but a few of those who have been proponents of conservation.

Municipal solid waste was not a major public issue until the affluent post-World War II period, when the rise of consumerism created a new phenomenon in the history of civilization: what social critic Vance Packard termed "the waste makers." In his 1960 book by that title, Packard wrote:

> *Man throughout recorded history has struggled—often against appalling odds—to cope with material scarcity. Today there has been a massive breakthrough. The great challenge in the United States, and soon in Western Europe, is to cope with the threatened overabundance of the staples and amenities and frills of life."* [1]

Suddenly we had become victims of our own success, wasteful and extravagant of natural resources. And, in Packard's view, industry was to blame in encouraging the consumer to adopt deliberately prodigal ways.

The waste problem was now being attacked from a number of angles. The anti-litter movement began gaining momentum in the 1960s, for aesthetic reasons (litter is ugly) as well as conservation reasons (litter threatens wild-life), along with public health reasons. At the same time, the idea of the "environment" as we know it today—an interconnected eco-system on a planetary scale—was beginning to take shape, driven in large part by works such as Rachel Carson's *Silent Spring*.

The earliest proponents of this global environmentalism focused on air and water pollution from manufacturing, symbolized by the dirty factory smokestack. From the mid-1970s, the focus shifted to hazardous waste, as epitomized by the Love Canal disaster. Hooker Chemical Corporation had been burying toxic wastes on the Love Canal site, near Niagara Falls, New York, from 1942 to 1953. But it was not until 1978 that a link was found between the toxic waste and the local residents' health, and the subsequent evacuation and relocation of the town's citizens became a powerful symbol for the emerging environmental movement.

Environmental activist groups have come a long way between the first Earth Day of 1970, and Earth Day 1990. Today they are well-organized, sophisticated in working with the press and with legislators, and more main-stream than ever. More than 25 percent of the American public provides financial or other direct support to environmental groups.

Their approaches vary considerably. Some, like Earth First!, promote civil disobedience to promote their goals. Others are more attuned to work-ing with established institutions. For example, the National Wildlife Feder-ation hosts a Corporate Conservation Council with fifteen member corporations that are deemed environmentally responsible.

Some other well-known environmental groups with an interest in the municipal solid waste issue, along with their stated charters, include:

- *Environmental Action Foundation.* Focusing on municipal solid waste, recycling, toxic substances, and other environmental issues, EAF is a national political lobbying association with 20,000 members and a $1.3 million budget.

- *Environmental Defense Fund.* With a staff of twenty-two and a budget of $14 million, EDF is a public interest organization of lawyers, scien-tists, and economists dedicated to protecting environmental quality and public health through public policy, legal action, legislation, and public education.

- *Greenpeace United States.* With 1.5 million supporters and a staff of 150, Greenpeace initiates active but nonviolent measures to aid endangered species and monitors conditions of environmental concern, including the greenhouse effect.

- *National Audubon Society.* Originally a wildlife conservation group, Audubon is a 500,000-member association that lobbies on such issues as recycling, energy conservation, and toxic wastes, as well as the protection of endangered species.

In many ways, Europe is far ahead of the United States in terms of environmental political movements, or "Green" parties, and U.S. industry can learn much from observing the European scene (see Chapter 5).

Legislators

Most of the key players in the waste management arena seek to influence legislation. The federal role in municipal solid waste management has evolved from a nonregulatory one prior to 1980 to a more regulatory one since 1980.

The packaging industry is no stranger to federal regulation. For example, the Department of Agriculture has regulated the packaging and labeling of meat and poultry in a series of acts beginning with the Meat Inspection Act of 1906. The Department of Health and Human Services, including the Food and Drug Administration, enforces regulations concerning packaging and labeling in foods, cosmetics, and pharmaceuticals. The Bureau of Customs enforces packaging regulations regarding imports. The Interstate Commerce Commission regulates containers transporting dangerous articles. And the Federal Trade Commission enforces regulations concerning deceptive packaging and labeling.

Prior to the 1969 National Environmental Protection Act, which established the federal Environmental Protection Agency (EPA), most environmental regulations were piecemeal and uncoordinated among several federal agencies. It was left to state and local authorities to make sense of them and add continuity. The target of these early regulations was pollution in its most visible forms: soot and smoke in the air, dyes and sewage in water, and litter on our roadsides. State and local authorities would eventually take the lead in environmental legislation, as in mandatory deposit laws (MDLs), and federal legislators would take their cue from the states.

A strategy supporting a waste management hierarchy has been in effect since 1976, when the EPA issued a position statement entitled "Effective Hazardous Waste Management (Non-Radioactive)." Under that hierarchy, source reduction first and recycling second are the preferred options for managing municipal solid waste. Combustion and landfilling are to be used only when the preferred options are unavailable or insufficient. The EPA recognized, however, that strict adherence to this hierarchy was inappropriate; Nevada, for example, would likely choose a very different mix of solutions from New York.

A major turning point in the federal role was the passage of the Resource Conservation and Recovery Act (RCRA) of 1976, which added enforcement to the Solid Waste Disposal Act of 1965. Although only part of the RCRA focused on municipal solid waste, the law gave equal importance to both municipal solid waste and hazardous waste. The EPA saw that the states were not implementing the law's waste disposal guidelines consistently; therefore, the EPA imposed minimum performance standards for municipal solid waste disposal facilities. These criteria were designed to set a floor standard for the states, many of which already had higher disposal standards in place, although others had no standards at all.

Under RCRA, the EPA effectively delegated enforcement authority to the states. In fact, the EPA hierarchy of source reduction, recycling, combustion, and landfilling was largely inspired by earlier state initiatives.

One area that has garnered increasing EPA interest is environmental product labeling. This would create a "seal of approval" for products that satisfy certain environmental criteria. California, Rhode Island, and New York were the first states to adopt labeling regulations that define "recycled," "recyclable," and "reusable," and state attorneys general met in November 1990 to propose a national labeling program. The EPA is observing these initiatives, and the FTC is reviewing an industry petition that would have the FTC develop guidelines for environmental labeling.

The EPA has also studied other countries' labeling programs, the first of which was West Germany's Blue Angel seal. Examples of products that carry the seal are adhesives and paints without solvents, aerosol sprays containing no CFCs, phosphate-free detergents, and inks and pigments that contain no heavy metals. Canada, Norway, Sweden, and Japan have adopted similar labeling programs, and the United S tates may have a program of its own before the decade has passed. Some of these labels and symbols are shown in Figure 1-2.

In 1991 and 1992, RCRA will be examined once again. Hundreds of proposed bills, including a national mandatory deposit law, will be debated during the process of reauthorizing RCRA. From this debate, a legislative mandate for a national solid waste strategy will ultimately emerge.

State and Local Initiatives

Businesses must contend with widely varying standards in different localities. While it is likely that some recycling legislation will be passed on a

Figure 1-2. Environmental seals of approval.

Green Cross   Green Seal

Germany's Blue Angel   Canada's Environmental Choice

national level, many states will continue to enforce higher standards, and policy implementation will be very much a local matter (see Figure 1-3).

Recent years have seen a flurry of legislative activity at the state and local level, especially regarding recycling. In the first five months of 1990 alone, sixty-five recycling laws were enacted in twenty-seven states. At least thirty states, plus the District of Columbia, now have comprehensive recycling laws, which require detailed statewide plans and/or separation of recyclables and contain one or more other provisions to stimulate recycling.

While the EPA set a national recycling goal of 25 percent by 1992, many states have adopted much more ambitious goals. Maine requires 50 percent recycling by 1994, and Washington 50 percent by 1995. Indiana set a goal of reducing the amount of waste landfilled and incinerated 35 percent by January 1, 1996, and 50 percent by January 1, 2001. Massachusetts expects to reduce its need for disposal by 56 percent by the year 2000 through source reduction and recycling. Wisconsin updated its 1983 recycling laws to mandate separation of recyclables from the waste stream and to ban them from landfills and incineration, including aluminum cans, cardboard, polystyrene foam, and containers made from glass, plastic, and steel.

Florida imposes a tax of one penny on containers made from aluminum, metal, glass, plastic, or plastic-coated paper if the products do not meet a 50 percent recycling rate by 1992, and a two-cents tax if this rate is not achieved by 1995. The tax is refundable if the container is brought to a drop-off recycling center.

Figure 1-3. State recycling goals.

State	Year Enacted	Goal	State	Year Enacted	Goal
California	1989	50% by 2000	Michigan	1988	50% by 2005
Connecticut	1987	25% by 1991	Minnesota	1989	25% by 1993
D.C.	1989	45% by 1994	New Jersey	1987	25% by 1992
Florida	1988	30% by 1994	New York	1988	50% by 1997
Illinois	1988	25%	North Carolina	1989	25% by 1993
Indiana	1990	50% by 2000	Ohio	1988	25% by 1994
Iowa	1989	50% by 2000	Pennsylvania	1988	25% by 1997
Louisiana	1989	25% by 1992	Rhode Island	1986	Maximum
Maine	1989	50% by 1994	Vermont	1987	50% by 2000
Maryland	1988	20% by 1994	Washington	1989	50% by 1995
Massachusetts	1987	20% by 1992	West Virginia	1989	30% by 2000

Another interesting twist to recycling may be seen in New York, where mandatory deposit laws (MDL) have been in effect since 1983. While the neighboring MDL state of Massachusetts has achieved a 90 percent return rate, New York's return rate for these containers is only 25 percent, and the state's beverage companies are allowed to keep the five-cent deposit per container if it is not returned. According to a spokesperson at New York's EPA Region 2, the result has been a $500 million windfall to the beverage companies. The state has asked for legislation to require the use of this unclaimed money to fund municipal solid waste management solutions.

Market development for recyclables was a key theme of 1990 laws. A recent glut of old newspapers in many parts of the country demonstrated the problem with collecting recyclables before finding markets for them. Among the approaches favored under the new laws were state government procurement programs mandating the purchase of recycled materials, tax credits for businesses that manufacture products from secondary materials, requirements that industries use recycled materials, and implementation of waste reduction and recycling programs at businesses. For example, an Indiana law gives priority for economic development grants to businesses that participate in the recycling infrastructure. Virginia established tax credits for manufacturers that use recycled materials. Iowa offers sales tax exemptions. Connecticut, California, Maryland, Wisconsin, and Missouri have all passed laws requiring newspaper companies to buy recycled newsprint. In at least seven other states, the newspaper industry has negotiated voluntary purchase and use standards. Wisconsin also requires plastic containers to be made from 10 percent reclaimed materials by weight.

Waste reduction has been an important goal of state initiatives, although creating policies to implement it has proved challenging. Strategies have included banning high-volume wastes and banning or taxing materials that are not being recycled. Eleven states and the District of Columbia ban yard wastes from disposal facilities. While some states collect these materials for composting, others ask homeowners to devise their own alternatives, such as leaving grass clippings on the lawn.

Since only half the municipal solid waste stream comes from residences, states are also requiring commercial businesses to separate their wastes. Many businesses already recycle because it saves them money. Of the 23 million tons of material recycled in 1988, more than half consisted of corrugated boxes, office paper, and lead-acid batteries recovered from businesses. Maine was the first state with a mandatory recycling law for businesses of fifteen employees or more. Rhode Island requires businesses to recycle glass food and beverage containers, aluminum, steel cans, newspapers, some plastics, office paper, and corrugated fiberboard; truck loads containing more than 20 percent of these materials are rejected at state landfills, and the commercial waste generators may be fined $500 for each violation

A number of states have been guided by model legislation drafted by the Coalition of Northeastern Governors (CONEG). This organization, which comprises the governors of nine states (the six New England states, Pennsylvania, New Jersey, and New York) as well as industry representatives and environmental groups, has provided leadership on a number of environmental initiatives.

More recently, the Source Reduction Council (SRC), established within CONEG in September 1989, focused on model toxics legislation, preferred packaging guidelines, and an informational program. Owing to a CONEG initiative, toxic material legislation banning the use of four heavy metals (lead, mercury, cadmium, and chromium) in packaging has been introduced in all CONEG states and four other states. In addition, the Packaging Standards Subcommittee made substantial progress on preferred packaging guidelines, including measurement methods for source reduction, consensus recommendations for package labeling, and voluntary standards for industry action on recycling, recycled content, and reusability. The Source Reduction Council's work during 1991 became bogged down in internal debates over its approval process and the participation of nonprofit groups. As a result, the fate of the SRC is uncertain.

Across the nation, states are helping to achieve a national recycling infrastructure, although their methods and approaches are inconsistent. In the interior region of the country, states tend to avoid broad separation mandates in favor of detailed planning requirements and disposal bans. In addition, most states have a long way to go before achieving their recycling goals. For example, Massachusetts, which set a 56 percent source reduction and recycling goal by the year 2000, had achieved only an 8–12 percent rate as of August 1990.

Retailers

A study by the Food Marketing Institute found that half of shoppers surveyed were willing to switch to a supermarket that promoted environmentally friendly products and packaging. Recognizing that some consumers will make the switch, some retailers have already put into place policies to capitalize on this trend.

WalMart, the nation's largest discount chain, with 1,400 stores in twenty-eight states, has asked its 7,000 suppliers to develop environmentally friendly products that the company would label and market as such. Loblaws, a Canada-based retail chain, has a full line of "green" products.

Seventh Generation, a Vermont-based mail-order company, is one of a new breed of catalog sales companies focusing exclusively on environmental products, from office supplies to clothing. Its catalog pages provide brief "question and answer" columns about such issues as composting, recycled paper, disposable diapers, and lawn care; to its credit, the answers are neither simplistic nor self-serving. For further information on each issue, the customer is provided with phone numbers and addresses of environmental organizations.

To assist in identifying "green" products, two independent organizations are offering environmental seals of approval. Green Cross, a division of California-based Scientific Certification Systems, has certified state-of-the-art recycled content and biodegradability claims for more than 350 products. All of these products are eligible to carry the Green Cross emblem and a specific explanation of the claim that has been certified, e.g., "this glass jar has been independently certified to contain 50% recycled glass." Green Cross has worked with leading manufacturers such as the Clorox Co., Fox River Paper Company, Owens-Brockway, and Wellman, and retailers including ABCO Markets, Fred Meyer, Raley's and Ralphs Grocery Company. In addition to certifying specific claims, Green Cross is conducting for manufacturers examinations of the full environmental impact of their manufacturing process. Another group, Green Seal, Inc., is based in Washington, D.C., and led by Earth Day 1990 Chairman Denis Hayes. The Green Seal will be applied to products that are judged to reduce waste, from manufacturing to disposal (see Figure 1-2).

Most of the "green retailing" is occurring not in the national chains but at the local or regional level. Health food stores are an obvious home for green products. One of the largest health food retailers on the East Coast, the Bread & Circus supermarket chain, has taken the stand of endorsing the aseptic juice box as environmentally sound packaging—at a time when its home state, Massachusetts, was considering a ban on the boxes, and after Maine had already banned them outright. In addition to providing pamphlets about the energy savings and waste reduction benefits of the package, the store provides collection bins for used boxes—offering a powerful example of what one retailer can do to influence the consumers' environmental choice.

Retailers in states with mandatory deposit laws, especially supermarkets, have found themselves involved in the recycling business whether they like it or not. Stores that sell beverages are often required by law to collect the containers and refund the deposits. By expanding their role as community recyclers, some stores have uncovered a marketing opportunity in what was initially seen as an inconvenience. Many offer to take back grocery sacks for recycling. Others help the consumer identify products on their shelves that have environmental benefits.

Retailers have also discovered that recycling their own waste is good business practice. Major supermarkets have been compacting their own used corrugated boxes for years, selling the material for recycling. This not only eliminates the disposal costs for brown boxes; it also creates a new source of revenue.

Manufacturers are likely to see increasing pressure from the retail outlet for products that satisfy the consumer's growing interest in environmental responsibility.

Packaging Industry

The typical corporate posture of yesterday was characterized by an unwillingness to take environmentalism seriously. If corporations did adopt "green" business practices, they were reactive in nature and pandered to the consumer's most superficial perceptions. Environmentalists and their value systems were seen in negative terms: anti-growth, anti-technology, and generally regressive.

We are now in a transitional period. Proactively "green" corporations are already identifiable, and they compete with each other for environmental leadership. At the same time, reactive companies have found themselves increasingly vulnerable to public rebuke. The rapid expansion of global ecosystem understanding and the evolution of infrastructure technology has paved the way for a meaningful corporate environmentalism.

The corporate environmental leaders of tomorrow will be those that help define and work within a philosophy of "industrial ecology." Their challenge is to develop new management tools and to make significant changes in products and packaging so as to have the gentlest effect on the natural environment. Indeed, environmentalism must eventually be fully integrated into the corporate culture.

As we have also noted, most American consumers surveyed in polls say they are influenced by a company's reputation on environmental issues. However, industry will be well-advised not to rely too heavily on consumer attitude surveys, which are often better measures of consumer intentions than of consumer practices. Industry should help define a rational policy agenda and therefore public response, rather than merely react to the pollsters.

By working more closely with the media and environmental groups, industry can exert a more positive influence on the consumer. However, historical antagonism may persist between environmentalists and corporations. Many companies lack the necessary skills to educate the media and many executives feel that they have been ill-treated by the media. The result is a vicious cycle of hostility to the press, more negative publicity, and continued antagonism.

Perhaps the greatest immediate opportunity for industry leadership lies in adopting life-cycle thinking in its environmental policy. This means exploring the full environmental implications of a product or process, from product design to raw material acquisition to energy consumption to pollution and hazardous wastes arising from the manufacturing process, and finally to post-consumer disposal options. (Chapter 6 explores this approach in greater detail.)

While life cycle analyses will not provide the definitive answers, industry can insist that the single-issue focus be replaced by broader criteria in legislation affecting its products and processes. It can demonstrate life cycle thinking in its environmental policies, and it can ask that legislators, the media, retailers, and consumers judge it by the same broad, balanced standards. In so doing, industry can expand the terms of debate and offer a powerful educational tool. It can help shatter simplistic dogmas and create a better understanding of the complexity of environmental issues.

Chapters 2 through 4 describe the structure of the packaging industry in greater detail, along with the constraints associated with different packaging materials and waste management infrastructures. Subsequent chapters detail how companies can get their environmental house in order while gaining consensus with consumers and environmentalists alike. Finally, guidelines are offered for working effectively in a rapidly changing socioeconomic climate even as the rules of the game are constantly revised.

The packaging industry has often adopted adversarial relationships with the other key players in the environmental arena. For years, the industry has sought to outrun or defeat regulations after environmentalists and others have influenced the political process. It can no longer afford to do so. Industry must build upon and strengthen its relationships with the key players in the process, and work within this loose partnership to define a new industrial ecology.

Notes

1. Vance Packard, *The Waste Makers* (New York: Simon & Schuster/ Pocket Books, 1960), p. 6.

The Packaging Industry

2

The packaging industry, as we have said, defies strict definition. At best it can be described as an aggregate of supply side and demand side materials, technologies, establishments, and institutions whose success or failure is directly determined by the ultimate consumer of packaged products.

The packaging industry is more than a single industry. It is both a large and intricate set of industries in itself, as well as a key determinant of success in the manufacturing, marketing, and distribution activities of many other industries. The supply side includes industries extracting base materials such as wood, metal ore, and sand, as well as the companies that process those materials: paper mills, aluminum processors, petrochemical companies, glassmakers. These businesses are typically characterized by high industry concentration with capital investment and specialized practitioner art as barriers to entry.

Between the base-material industries and the packer are multiple layers of converter establishments, which vary widely in terms of the number of layers and the degree of value added. For example, a maker of specialty laminated films may use polymers or resins from several sources, which have been extruded as film or sheet stock by specialty processors, then combined by a single specialized laminator and coated or metalized with materials from other specialty converters.

The packers—the producers of packaged goods—belong to the demand side of the industry. They are characterized by their target consumers and by the preservation technologies and marketing strategies required in their marketplace. Major categories are consumer packaging, institutional packaging, industrial packaging, and military preservation packaging.

The demand side of the industry also embraces the technologies of distribution through to the integrated retailer/consumer environment, where the package becomes a significant force in establishing market share among competitive products. In its broader sense, the packaging industry also includes retailers and consumers—both household consumers and end-user industries. It also includes waste management companies, as these businesses return post-consumer packaging as a resource back into the supply chain.

Several industry estimates put the value of the U.S. packaging industry at $65 to 70 billion. These estimates usually include products used for packaging and related items: rigid and flexible containers, wrappers, components like caps and closures, and labels, as well as tapes and twines. The estimates do not include institutional and retail bags, wraps, trays, household totes, single-service cups, and material handling containers, which would easily push the estimates to well over $70 billion. Nor do they include all of the businesses that influence, or are influenced by, packaging.

In short, packaging is a vital contributor to the economy at both a national and a global level. Nevertheless, to the majority of American consumers, packaging is very much an invisible commodity, an irrelevance. When packaging does become an issue—as in the case of aerosol sprays, or the single-serving juice box—there is a tendency to focus on single, isolated issues and to seek simple answers, giving little thought to the broad implications of those narrow solutions.

Behind every package there is a wide and complex universe of industries and issues that must be addressed before constructive change can take place. This chapter will outline the structure of the packaging industry, so that all the key players—activists, consumers, legislators, the media, educators, and businesses—can better understand the actions that must take place throughout the system in order to bring about environmental progress. Demands for single-issue solutions must be replaced by systemic thinking, and by systemic solutions.

Structure of the Industry

Any picture of the packaging industry is subject to a virtually infinite number of possible renderings. If one starts with the raw material, the path that material takes from source to packaged product can vary greatly, depending on the material, the nature of the product to be packaged, and its market.

Perhaps the simplest picture we might draw of the packaging industry as a whole appears in Figure 2-1. This tree chart of the industry is intended as a broad structural outline; in reality, businesses rarely fall neatly into these categories. There may be a great deal of overlap, integration, or consolidation of several functions in one business. Glass packaging, for example, is produced by companies that both melt the raw materials and form them into glass containers. By contrast, the flow of materials in the aluminum industry involves more types of businesses—mining, primary mills, and drawing and rolling mills—before the material goes to separate establishments spe-

Figure 2-1. Structure of the packaging industry.

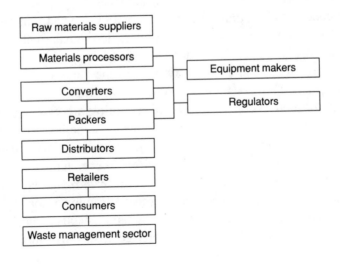

cializing in fabricating cans, pails, foils, tubes, closures, semirigid containers, or other packaging.

The packaging industry starts with the supply of resources, the mining or gathering of the *raw materials*, such as wood for paper, sand for glass, bauxite, and other ores for aluminum packaging, and crude oil and natural gas for plastics. Industries at this level include logging, petroleum refining and natural gas processing, and metal and mineral mining; they are the first link in the chain.

The *materials processing* layer is made up of the various facilities that turn the raw resource into materials such as paper, metal, plastics, and glass. Of course, not all of these materials go into packaging. Plastics, for example, consume less than 2 percent of the total petroleum resource used in the United States, and plastic packaging makes up an even smaller percentage. The industries at this level are pulp and paper mills, petrochemical plants (specifically, those involving plastic materials and synthetic resins), steel foundries, aluminum mills, and glass container companies.

The *converters* are a broad set of companies, both large and small, that enhance the basic materials and combine them with other materials. They are specialty businesses that provide custom materials on demand, for example, corrugated fiberboard boxes, folding cartons, aluminum foil laminated to paper, or composites of several different plastic films. This industry also includes coatings, adhesives, caps, and labels.

Conversion can be a multilevel process, with value added at each level. Converters may or may not be vertically integrated with the material suppliers. In the supply of container board and industrial papers, for example, we find vertical integration all the way back to the natural resource; forest management and reforestation is also part of the chain.

In addition, conversion can take place in multiple layers of organizations or in a single organization. To cite again the containerboard example, the corrugating plant receives raw materials in roll form from the paper mill, which may contain a high percentage of virgin fiber material for the exterior layers of the box and recycled fibers for the corrugated inside layer, and then combines them into corrugated containerboard. Printing or coating, cutting, scoring, folding, and seaming is done at the same establishment. The box is then delivered to the packer, where it is erected, loaded with the product it is designed to protect, sealed, and labeled for delivery. This is an example of vertical integration with multiple stages.

An example without vertical integration is the supply of complex composite materials that may be coated and printed with specialty printing. Here there may be several components that when combined provide the product protection, product compatibility, and other required characteristics.

For example, there may be a need for a coated paper stock with a high-barrier characteristic to preserve a sensitive product. The paper would come from the paper mill, made from high-quality virgin wood pulp. It might also be combined with a certain amount of reclaimed fiber that would come to the mill from the waste management infrastructure. Then it could be supplied to a converter that might apply a moisture barrier coating to the sheet, perhaps a polymer coating from another material source. Foil might also be laminated to that sheet as a gas barrier, supplied by an aluminum processor. At this stage the material might go to a high-quality printer. Consequently, there may be multiple converting stages in simply producing the packaging stock, well before it is ever sent to a packer to be formed, filled, and sealed.

Packaging therefore involves a very complex supply chain. Each succeeding process step in preparing the material or package form adds value before it ever reaches the packer. And at each level, choices of materials and processes are based on multiple levels of concern. The package and its materials must meet functional expectations, such as containment, preservation, product compatibility, communication, and structural strength, at a reasonable cost. There are also promotional aspects: the attractiveness of the package, its convenience, and its ability to sell the product. At the same time, the packaging company has to deal with federal, state, and local regulations: from EPA rules on water contamination, air emissions, and hazardous waste, to Occupational Safety and Health Administration regulations that protect workers dealing with heavy equipment and potentially harmful materials, to Food and Drug Administration requirements that protect the health and safety of the consumer.

Equipment makers, which supply the converters and packers, are also regulated. Complex rules will govern what can be achieved along this supply chain. Major capital investment in packaging or process equipment innovations are often required to achieve the desired level of performance at acceptable cost.

Demand Side of Industry

Any one of the components in our tree structure of the industry has a supply and demand side. However, within our conceptual framework, *packers*—the producers of packaged products—belong to the demand side. The packer is the first one to define the actual package at the consumer end, although many other players further upstream will also influence the end product.

There are numerous examples of strategic alliances and integration in packaging. A converter may decide to integrate forward or backward with a proprietary system or a piece of equipment, in order to secure a captive market. An example is Tetra Pak, originator of the ubiquitous drink box. With headquarters in Lausanne, Switzerland, Tetra Pak is a converter that developed a package, a preservation process, and equipment, all in one. What Tetra Pak developed is licensed in the United States as a complete system. Manufacturing the package means using Tetra Pak's equipment, process, and materials. There are competitive systems, but if the packer wants the Tetra Pak box, it does not have many options to look elsewhere.

Packers commonly integrate backward, to capture their own source of supply. Major brewers often have their own can making operations. Some have integrated forward as well; Anheuser Busch, for example, has its own waste management infrastructure—a system for reclaiming aluminum cans. A packer may even be integrated forward to the retail outlet. 7-Eleven, like many retailers, integrated with a packaging company to offer its own proprietary brands.

Foods and beverages represent the largest user of packaging, with $24.1 billion in packaging expenditures in 1989 (see Figure 2-2). Food packers run the gamut from canners to frozen foods packers. Packers may be intensely specialized or may span a spectrum of food and beverage products: processed meats and poultry, seafoods, processed fruits and vegetables, dairy products, cereals, bakery products, and confections. Specialization is typical in some products: Beverage people are often in the beverage business only, while snacks producers usually specialize in snacks. However, the snacks category has become intermingled with more nutritional varieties that border on staple foods.

Another large user of packaging is the pharmaceutical industry. Here we can identify both manufacturers of ethical (prescription) drugs and makers

Figure 2-2. Top ten companies in package and packaging material expenditures, 1990.

1	Philip Morris	New York, NY	$2,568,450*
2	Anheuser Busch	St. Louis, MO	$2,300,000
3	PepsiCo	Purchase, NY	$1,460,880*
4	Procter & Gamble	Cincinnati, OH	$1,385,500*
5	Coca-Cola	Atlanta, GA	$1,161,720*
6	Coca-Cola Enterprises	Atlanta, GA	$1,086,932*
7	RJR Nabisco	New York, NY	$ 908,800*
8	Seagram USA	New York, NY	$ 712,920*
9	Sara Lee	Chicago, IL	$ 575,000
10	Unilever US	Greenwich, CT	$ 519,820*

* *"Packaging Magazine" estimates*

of over-the-counter (OTC) drugs in the areas of biological, medicinal, botanical, and pharmaceutical preparations. Packaging expenditures in these areas totaled $6.6 billion in 1989. Overlapping those sectors are medical products, including dental equipment. Then there is the diagnostics business, which supplies both chemical reagents and equipment.

In the *distribution* business, we move into the area of industrial packaging. Distribution requires packaging for the safe and efficient movement of goods from one point to another. Industrial packaging has to both protect the product and protect the rest of the freight from the product. It has to meet industry-standard handling requirements. And it has to add value in terms of reducing the costs of handling and distribution to the consumer.

Then there are categories within industrial packaging. For example, a refrigerator is shipped in an industrial package that has no selling function; its purpose is to protect the product during handling and shipping. In computer products, mainframes are also shipped in industrial packaging not intended to "sell" the product; the package may be printed with a series of logos and company identifiers, but it imparts little information except for handling and distribution. However, the shipping of personal computers represents a blend of both industrial and consumer packaging. The packaging must facilitate handling and minimize distribution costs, but it also carries a marketing message, because the consumer will personally handle and unpack the product.

The distribution business can be owned and managed by the packing company, or by private industry invested only in distribution and/or private warehouse operations. A great deal of forward or backward integration can be achieved. Many retail outfits have their own distribution facilities. Suppliers deliver to their warehouses, and the retailers take it from there. This has been a long-term, well-established practice to control delivery costs and to take advantage of buying practices.

Some packaging functions are critical in distribution. Protection of the product from both dynamic and static forces is primary. Dynamic forces are the shock and vibration of transit itself, coming from trains, trucks, ships or aircraft, and in devices like forklifts and hand trucks, as well as in manual handling. Static damage takes place when the product is not in transit; it includes top-to-bottom compression and other damage from stacking.

What is often considered overpackaging is, in fact, performing a needed function: reducing overall cost to the consumer by minimizing damage and product recalls. Computer-based simulation techniques have been applied to anticipate the demands that will be placed on industrial packaging in virtually any type of distribution. Industrial packaging has truly achieved a high degree of efficiency, minimizing the use of packaging materials.

At the *retailer level*, consumer packaging is used principally to sell the product and deliver product claims. That is, the retail package does the point-of-purchase sales job by displaying information about the product and describing features that distinguish the product from its competitors while protecting the product quality until consumed. The package is designed to stand out on a retail shelf that is crowded with competitive products. A factor contributing to complex packaging is convenience. Single-serving foods, portion packs, and microwavable containers generally increase the volume and complexity of packaging. Depending on how marketing interprets the sales function of the package, there can be multiple layers of dissimilar material that intensify the burden on the waste management infrastructure.

Other elements in multiple-layer packaging are necessary for the protection of the consumer. Any consumable item needs protection from spoilage or contamination. Additional layers are used to supplement more permeable packaging or to provide seal integrity. Further layers and devices are used to protect the consumer from malicious tampering or to provide child resistance. There is little latitude allowed the packers at the retail level when they need to consider such factors as regulation, preservation, safety, and the physical requirements of the display environment.

One simple and fondly recalled example of value-added packaging that serves a waste minimization function is the jelly jar of the 1950s that doubled as a child's drinking glass; another is the peanut butter jar that is calibrated to serve as a measuring cup. These packages were designed to influence their final use. Without consciously thinking of it as an environmental decision, many households have for years purchased products in the large economy-size package to refill the smaller and more convenient dispenser package at home.

Consumer package design often goes beyond the containment, preservation, transport, and communications function of a package and has done much to enhance our current life-styles. Environmentally responsible marketing considers post-consumer waste management as an additional functional consideration for consumer package design.

As this decade progresses, the term "waste management" may prove to be something of a misnomer. To achieve a true industrial ecosystem, there is no room for the very concept of waste. At the post-consumer stage, all materials could find a useful life someplace else, either through reuse, recycling, or conversion to energy with little left for disposal as waste. At the end of the packaging chain, there needs to be a new approach to resource recovery and return.

The Impact of Environmental Decisions

Any environmental decision has to be attuned to a myriad of performance, promotional, and regulatory considerations. The industry and the issues it faces are immensely complex, and the environmental solutions are complex as well. A solution may appear simple from the consumer's perspective, but when it is applied to a whole system of this complexity, the problems envisioned may be nearly unresolvable. For this reason, the packaging industry itself is perhaps best positioned to take a leadership role in resolving environmental issues. Few observers outside the packaging industry have made the investment in time and education to understand the complexity of the system.

The packaging industry has traditionally been concerned with the fate of its end product, at every step of the supply chain. Even at the base material level, innovative companies develop concepts and materials that will improve retail performance, or satisfy a consumer need, and they use "pull through" to persuade packers to use them. The aluminum industry is a good example; its innovations in can fabrication technology, the recycling infrastructure it established, and its promotional efforts are combined to sell the aluminum can to the consumer. However, innovation has limits and is controlled, in addition to free market controls, by regulators who take into consideration how the product performs with the consumer. No package can sell unless it achieves a desirable level of safety and delivers its claims.

Quote → The *consumer* level is where the package performs its intended function, and where the environmental impact of the package is ultimately determined. A package designed for recycling cannot fulfill its mission unless an accessible waste management infrastructure exists and the consumer separates and returns the package to the system through this infrastructure. While the packaging industry can design its products to influence and instruct the consumer, the decision making remains in the consumer's hands.

One of the basic premises of this book is that the role of the responsible consumer is critical. At the same time, there is a gap between the consumer's environmental concern and actual buying practice. As we have seen, consumers overwhelmingly indicate their endorsement of environmentally responsible products in consumer attitude surveys. In actual practice, however, consumer behavior has been slow to change.

While much of it is unseen by the consumer, the packaging industry practices environmentalism in a number of forms. Produce packaging is an example. Much of the inedible and otherwise wasted portion of produce is removed at the source by the packer. If lettuce were shipped as it was produced, with no stripping or paring of its outside leaves, there would be 40-60 percent more waste that would add to shipping costs and that the consumer would have to dispose of. Those leaves are also an excellent source of composting which, when removed at the source, is very close to where it can be used most efficiently.

Recycling of factory scrap, lightweighting of plastic bottles, reuse of wooden pallets, bulk shipment of commodity goods, and numerous other industry practices are efficient uses of resources that also reduce costs. Indeed, much of what has been achieved as cost improvement in the packaging industry has produced environmental improvement.

Much more needs to be done, and change can be inspired at any level. But no change can be imposed on the product or package without affecting an enormous number of decisions throughout the system. The debate on environmentally sound packaging has historically focused on waste management, with selected recommendations for the choice of materials, but with few constructive suggestions directed at decisions higher up the supply chain. Demands cannot be made on the system arbitrarily and without forethought. There are technical, economic, and regulatory requirements in this matrix that have to be addressed.

Before making sweeping judgments such as "all plastics are bad," people should consider the example of the "castle crate" (see Figure 2-3). This is the

Figure 2-3. The castle crate.

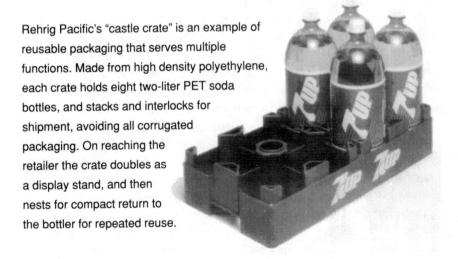

Rehrig Pacific's "castle crate" is an example of reusable packaging that serves multiple functions. Made from high density polyethylene, each crate holds eight two-liter PET soda bottles, and stacks and interlocks for shipment, avoiding all corrugated packaging. On reaching the retailer the crate doubles as a display stand, and then nests for compact return to the bottler for repeated reuse.

reusable, stackable, returnable plastic crate used to deliver large soft drink bottles from packer to retail store. It serves as both industrial packaging and a display case. The castle crate offers a very good model of environmental soundness. For one, the return cycle is built in; the truck that carries the new bottles to the store can take back the empty crates. More importantly, the crate has been engineered for multiple reuses.

Indeed, the much older example of the wooden pallet has supported this thinking for decades. More than forty years ago, the Grocery Manufacturers Association (GMA) put in place a standard hardwood pallet system with multiple reuse in mind. GMA standardized the size of the pallet, the way in which it was handled, and structural considerations such as types of wood, nailing patterns, and so on. The pallet was designed to move through the system and back again several times. The GMA pallet size is now standard in the United States. A similar system based on metric dimensions, but close in size to the U.S. version, is used in Europe.

This is another form of source reduction. Source reduction is traditionally thought of as "lightweighting," reducing the packaging material to its minimum functional limits. An alternative mode of thinking is deliberately to extend the lifetime of the package; while this would increase the initial materials input, it would promote multiple use and reduce the total packaging required for the volume of product shipped. The typical life expectancy of some current packaging forms is shown in Figure 2-4.

Reusable packaging is a concept that is now being discussed and promoted as an environmental solution. It works—indeed, it has worked successfully for decades. However, in the face of this complex industry structure, reusable and returnable packaging concepts must be thought through in a systemic manner within the infrastructure, and often the consumer is the most critical component of the system.

Environmentalists are now active participants at the consumer level of the industry as well as with legislatures. Environmentalists can be more effective if they understand the full structure of the packaging industry, where the major obstacles stand and thus, how they can achieve their agendas. Indeed, even within the industry, not all the players know the full complexity of the system. The packaging industry needs a better understanding of what environmental-minded consumers are seeking, and are willing to support, to be better equipped to initiate successful environmental responses.

Figure 2-4. Useful lifetime of various packages.

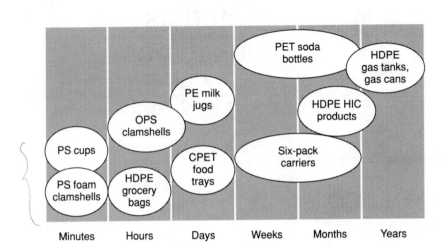

This is not an impossible task. There are complexities to be understood as prerequisites to change, but they are not insurmountable barriers. There have been people and organizations committed to understanding the full problem, and they have done so with some measure of success. Procter & Gamble is an excellent example of a company that has effected meaningful change throughout the packaging industry structure, from package design to materials acquisition to creating a waste management infrastructure. The alliance of Du Pont and Waste Management, Inc., to create a plastics recycling infrastructure is one of Du Pont's many systemic approaches. (See case studies of Du Pont and P&G in Chapters 8 and 9.)

Enormous opportunities remain for industry, activists, and informed consumers to achieve environmental progress through systemic, long-term solutions.

3

Materials: Limitations and Opportunities

The opportunities for environmental advances in packaging are as limitless and varied as the materials themselves. Beyond the most basic materials—paper, glass, aluminum, steel, and plastics of all kinds—there is a dizzying array of new composite materials—combinations of paper and aluminum, plastic and paper, and so on—that can be tailored to virtually any packaging specification.

While packaging materials and designs have traditionally been selected on such criteria as consumer appeal, convenience, utility, and cost, we are now beginning to see a new set of design and selection criteria: the environmental consequences of the manufacture, use, or post-use fate of the package. Since it is the material and the way it is constructed that determines the environmental impacts of the package, this chapter examines each packaging material in turn to provide a better understanding of the environmental issues surrounding their production, their transformation into packaging, and their disposal.

Paper

Paper, the most familiar and widely used packaging material, accounts for 48 percent of all packaging. Whether in the form of milk cartons, gift boxes, take-out food containers, or heavy corrugated cartons, packaging accounts for more than half of the paper manufactured today. In 1990, shipments of paper packaging are estimated at $39 billion. And, in spite of losing market share to plastics, the use of paper as a packaging material overall continues to grow.

As a portion of all solid waste in landfills—including newspapers, office waste, and other paper products as well as packaging—paper accounts for 54 percent by weight and 47 percent by volume. Clearly, paper represents the largest opportunity for waste reduction and recycling in the packaging industry.

Paper already has a long history of recycling. In fact, the paper industry in the United States began as a recycling industry, converting old cloth fibers into paper. Paper in the past has been made from straw, hemp, cotton and flax, but wood became the material of choice with the development of chemical processes that would free up the fibers from the wood, beginning with the soda process developed in the 1850s.

In 1990, the U.S. paper industry used about 81 million tons of fiber, of which 26 percent was derived from wastepaper and 74 percent from wood (see Figure 3-1). Unlike many other regions of the world, North America has an abundant supply of wood suitable for making paper, and environmentally responsible paper companies have ensured a continuing supply by replacing clear-cutting practices with sound forest management programs. As a result, paper companies can now grow trees faster than they are cut. Pine plantations and well-managed tree farms are contributing to maintaining a proper ecological balance.

The Making of Paper

The basic papermaking process (see Figure 3-2) is relatively simple: a dilute suspension of fibers and water is introduced at the front end of the paper machine, the sheet is formed as water is drained off, additional water is

Figure 3-1. Sources of fiber as a percentage of overall fiber for paper production in the United States.

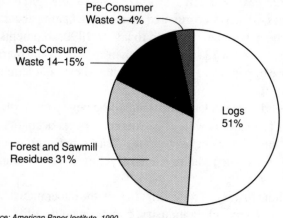

Pre-Consumer
Waste 3–4%

Post-Consumer
Waste 14–15%

Logs
51%

Forest and Sawmill
Residues 31%

Source: American Paper Institute, 1990

Figure 3-2. The papermaking process.

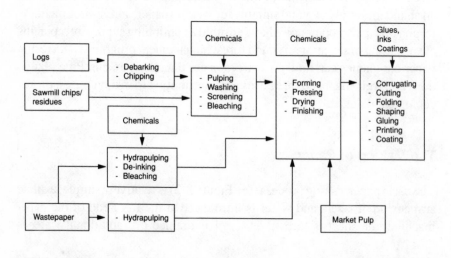

mechanically removed, and the sheet is dried to about 8 percent moisture. In essence, today's papermaking methods are not unlike those of the ancient Chinese, who first made paper as we know it in the second century. Paper-making equipment, however, has become more sophisticated and auto-mated, and fiber properties are now enhanced by the addition of a variety of chemicals—for example, to provide wet strength—or by the application of surface coatings for better abrasion resistance or printability.

To make paper, usually softwoods and hardwoods are combined. Wood for paper is supplied in the form of logs or residue chips from sawmills or plywood operations, with over 37 percent coming from manufacturing or forest residues.

The fiber material from which paper is made, called pulp, can be produced by chemical or mechanical means. The recovery of papermaking fiber depends on the pulping process used, about 45 percent from chemical and up to 95 percent from mechanical. About 61 percent of the fiber used by U.S. paper mills is produced by the chemical processing of wood, frequent-ly at the same site as where paper is made. Besides chemical processing, wood can simply be ground by metal disks or against stones to produce finely divided fibers, used for products where strength is not important, e.g., newsprint. There are also combinations of chemical and mechanical steps, such as a semichemical process called NSSC (neutral sulfite semi-chemical), which provides fiber for corrugating medium, the fluted middle layer of containerboard.

The most common chemical process is the sulfate or kraft process, which uses sodium sulfide and sodium hydroxide to remove the lignin that binds the cellulose fibers together in the natural state. The strong, brown fibers (pulp) produced at these mills are particularly suitable for bags, sacks, and linerboard (the outer surface of the familiar brown box). When the pulp is whitened by chemical bleaching, it can be used for such products as food containers, printing papers, and tissue.

Despite the relative simplicity of the process, building a papermaking facility is very capital-intensive—investments in the order of $200,000 per daily ton are not unusual. The average size of papermaking machines has increased; the annual capacity of a new kraft paperboard machine is in the range of 350,000 to 375,000 tons a year. However, a number of smaller, older units are still in operation, which provide a reliable supply of paper for packaging. In 1989 there were forty paper machines in the U.S. that were over forty years old producing packaging grades, many operating on wastepaper.

In addition, the cooking chemicals are expensive and must be recovered for both environmental and economic reasons. The recovery equipment for the chemicals is another major capital investment, and because the chemicals contain sulfur, the furnaces must meet strict atmospheric pollution controls. However, the economic benefits include the production of steam for energy, as well as the reuse of high-cost chemicals.

Although wastepaper has always been a good source of fiber for paper-making, it is becoming even more important. In 1990, about 22 million tons of wastepaper were recycled by U.S. paper mills, of which about 68 percent was used in packaging products. As recycling efforts intensify, wastepaper use should increase significantly—to an expected 30 million tons by 1995. To ensure a consistent supply of wastepaper, major producers of recycled containerboard and boxboard have some degree of backward integration to paper collection: the wastepaper must be sorted, cleaned, and defiberized; then it can be used either alone or mixed with virgin fiber.

Paper Packaging

Paper packaging can be conveniently divided into four groups, according to their use:

1. Corrugated and solid fiber boxes—the largest segment, with shipments estimated at $18.1 billion
2. Folding and setup boxes—estimated at $7.3 billion
3. Paper bags and sacks—estimated at $8.8 billion
4. Miscellaneous (e.g., fiber drums and cans)—estimated at $4.6 billion

Note that *papers* refers to the lighter-weight products made almost exclusively from bleached or unbleached chemical wood pulp. *Paperboard* is a heavier-weight product made from virgin or recycled fiber.

Corrugated and Solid Fiber Boxes

The familiar brown box is considered to be the best shipping container available for the price: it is strong; it provides protection; it can be marked for identification; and it can easily be recycled. It is made from container-board, a multilayer product: the outer facings are linerboard, and the inner layer is corrugated (or fluted).

The unbleached kraft paperboard used as the outer facing of corrugated or solid fiber boxes is made from at least 80 percent virgin fiber. Currently, recycled linerboard accounts for only 3–4 percent of U.S demand. Corrugating medium, the middle fluted layer, can be made either from semichemical pulp or wastepaper.

Although plastic is more durable, corrugated containers have easily retained market share. Industry sources estimate that United States consumption of corrugated containers is about 125 boxes per person. This is about 16 percent higher than in Japan and twice the European consumption.

Folding and Setup Boxes

Folding boxes and cartons account for about 94 percent of the boxboard market and include a wide range of products such as food and beverage containers, retail boxes, and toy and cosmetic boxes. Although there has been some erosion of market share by plastics, especially for milk and juice cartons, overall market demand has increased, especially for containers of convenience packaged foods and ready-to-cook meals.

Bleached paperboard is used primarily for milk cartons, food service products, ice cream and frozen foods, and high-quality packaging such as cosmetic boxes. It contains not less than 80 percent bleached virgin wood pulp. Recycled paperboard is made from a variety of wastepapers, which may be clay-coated to provide a better printing surface when used for folding cartons. Cereal boxes are an example of this process.

Paper Bags and Sacks

The market for grocery bags and shipping sacks has declined over the past fifteen years because of a dramatic shift away from paper to plastics. However, the ratio of paper to plastic seems to be stabilizing, with paper accounting for two-thirds of this market. The market for specialty bags—variety merchandise, produce, and household food storage bags—has also declined. Here plastic is clearly the dominant material.

Supermarkets are actively supporting the recycling of paper grocery sacks, often by providing special collection bins, in addition to encouraging customers to reuse the bags as many times as possible.

Fiber Cans, Drums, and Miscellaneous Packaging

This large commodity market includes (1) composite fiber cans used for frozen drink and juice concentrates, dry foods, and packaged mixes, and (2) fiber drums used to ship chemicals, detergents, resins, and parts and accessories. This market was adversely affected by the large market lost to plastics in the composite can business. However, industry sources see new end uses developing, which could repair some of the damage.

Environmental Issues in Paper Packaging

The consumer's general perception is that paper packaging is environmentally friendly because paper is biodegradable. If it is exposed to air and water, paper certainly will biodegrade, although wax and plastic coatings slow the process considerably. However, the paper that goes into landfills is largely sealed off from the elements and has little chance of degrading, even over many decades. Archaeological digs into landfills have turned up thirty-year-old newspapers that were still intact, and quite legible. Biodegradability becomes a more useful concept where roadside and marine litter are concerned, and will become more important as composting becomes a more widely available waste disposal option.

Consumers also favor paper packaging because it is recyclable, although, with the exception of corrugated boxes, very little paper packaging is actually recycled. However, paper packaging is often made from recycled wastepapers, of which the most important sources are corrugated boxes and old newspapers. Used corrugated boxes, which are easily separated from other kinds of papers, are the largest single source of wastepaper and are part of a well-established recycling infrastructure. In 1990 alone, almost 11 million tons of corrugated boxes, about half of U.S. production, were recycled by paperboard mills. However, household paper packaging waste has not been widely recycled for two reasons: (1) The collection infrastructure does not yet exist, and (2) the materials are not clearly defined. Household paper packaging may be made from virgin fibers, recycled newsprint, corrugated, or some combination of these, and inks and coatings make the sorting process even more problematic.

Wastepaper recycling, which increased during World War II to provide raw material in response to a very high demand for paperboard packaging, is once again receiving special attention because of environmental concerns. Recycling offers a major opportunity to conserve energy and water while preserving natural resources: For every ton of recycled paper used, an estimated ten to seventeen trees are spared.

The papermaking process is energy-intensive. Although pulp mills provide steam from the combustion of tree bark as well as spent chemicals, additional energy must be generated or purchased. Recycling paper translates into an average of 4,100 fewer kilowatt hours per ton over processing virgin fiber. Water savings amount to 50 percent of the costs of processing virgin fiber—unless de-inking is required, in which case water savings are reduced to about 15 percent. Finally, recycling can reduce water pollution by as much as 35 percent and air pollution by as much as 74 percent.

In addition to recycling, the paper industry has invested billions of dollars for process improvement to meet environmental regulations. For example, pressure to reduce discharge of chlorinated organic compounds has forced chemical pulp mills to look at ways to reduce or eliminate their use of elemental chlorine for bleaching, believed to be the prime culprit for the discharge. Substitution of chlorine by chlorine dioxide, oxygen, and hydrogen peroxide is being implemented at most bleached chemical pulp mills. Because the process is water-intensive, mills recirculate as much water as they can and have built their paper machines to mechanically remove as much water as possible. This has resulted in significant decreases in manufacturing costs.

The challenge for the mills is to move from a demand-driven to a supply-driven environment. Their major concern has always been to secure a consistent source of wastepaper of the proper quality and to avoid the price volatility that has historically characterized this demand-driven business.

As wastepaper supply increases because of recycling legislation, paper companies can look at opportunities to provide a more stable supply and pricing environment.

Paper companies are active recyclers, and most are exploring ways to use even more wastepaper—both mill scrap and post-consumer waste—without detracting from their products' strength, durability, and attractiveness. Certainly, the best market for wastepaper is the paper industry itself.

Some paper companies need to be convinced that the recycling movement is here to stay. Their facilities are large and efficient, with equipment designed for the use of virgin fiber. Many have large investments in mills balanced between wood pulping and papermaking that are not easily converted to accommodate the use of secondary fiber. Such changes can require significant investment.

For example, the installed cost of a state-of-the-art deinking system that can handle about 300 tons of waste paper a day is in the range of $8 million. In order to process the variety of inks, toners, dyes, and coatings found in office and printing papers, the washing systems traditionally used by U.S. mills are not sophisticated enough; flotation modules must be added and chemical formulations used. Mills that currently buy wood pulp rather than produce it have an opportunity to replace the high-cost purchased fiber with a de-inked pulp. For many companies this means adding a new or expanded disposal system for the effluent from the de-inking plant and increased sludge volume. This can be a problem for mills in urban areas with little available land.

To secure financing for recycling projects, mills must justify the investment to either their own corporate financial group or lending institutions. Responsible lenders must be sure that projects are financially sound. They need assurance about the long-term supply and price of wastepaper, the ability of the mill to compete on a cost and quality basis, the volume of the output that is committed and for what period, as well as the anticipated pricing policy.

Sourcing of waste paper is critical to the success of recycled paper mills. There is some infrastructure in place to collect and recycle corrugated boxes created out of necessity by mills requiring large volumes of wastepaper. However, securing a consistent supply of higher quality office papers is still in the development stage. Entrepreneurs are establishing a position by setting up office collection systems, organizing collection at the school level and trading services such as document shredding and office cleaning for paper.

Companies entering the recycled paper market must offer more than an environmentally friendly product. They must provide competitive pricing and consistently good quality, meet the packers' and carton makers' requirements for trouble-free operations, and be supported by technical service. For example, recycled paperboard must cut properly, accept printing, and run smoothly on the end-users' machines. Recycled paper producers must also demonstrate a commitment to the market through investment in quality equipment and by maintaining efficient facilities.

The challenge for the paper packaging industry is to use as much waste material as possible in packaging while still meeting the user's requirements for strength, protection, and other criteria. For example, the "Good News" grocery sack developed by Stone Container uses over 20 percent old newspapers without sacrificing quality and strength. To be successful over the long term, products and packaging made from wastepaper must compete with virgin fiber products. Short-term "fad" sales to exploit the green marketing trend will not secure long-term markets.

Glass

Glass is the second most common packaging material, accounting for 25 percent of all U.S. packaging. It is also one of the oldest packaging materials known. The Egyptians are believed to have used glass as early as 3000 B.C. And Pliny wrote in the first century of how Roman sailors discovered glassmaking: cargoes of soda burned in fireplaces on sandy beaches fused with the sand to form glass.

Today, glass is made from the same basic materials used for centuries: sand, soda ash, and limestone. Glass container production was first automated by Michael Owens in Toledo, Ohio, in 1903, but the basic principles have changed very little over time. The sand, soda ash, and limestone are mixed with crushed glass, called cullet, and fired in a furnace to about 2700 degrees Fahrenheit. From there the molten glass is cut with shears into a gob, which corresponds to one container. The gob goes down a chute to a blank mold where it is forced over a neck ring; there the threading or other finish of the bottleneck or jar rim is formed. The gob also acquires the general shape the bottle will ultimately have, although it is smaller since it is not yet filled out. The parison, as it is now called, is transferred to a final mold, where it is blown out to its finished form. The finished container is annealed, a process whereby the container's temperature is raised and systematically cooled to relieve any residual stresses in the product.

As a packaging material, glass holds many advantages: It is rigid, transparent, inert, impermeable, odorless, and microwavable. Consumers view glass as a premium packaging material. Glass is the material of choice for wines, fruits, sauces, condiments, and many other types of foods.

Container glass had witnessed a steady erosion of its market share in the 1980–1985 period, with the greatest loss in the beer sector. Once the largest customer of glass containers, the beer industry bought 120 million gross of glass containers in 1980, but over the following five years, billions of glass beer bottles were replaced by aluminum cans. Glass was also losing ground to plastics in foods and soft drinks markets, but at nowhere near the precipitous rate in the beer market. Shipments of glass containers dropped by 5.8 billion containers from 1980 to 1985, with 85 percent of that loss accounted for by breweries.

One reason why breweries converted to aluminum cans was the faster fill rate; that is, it is quicker to fill a can than a bottle. In addition, because of its greater volume and weight, glass is costlier to ship. The aluminum can represents a compact, efficient way to deliver beer.

Since 1986, however, container glass has been staging something of a comeback, growing at about 5 percent per year. Losses in the beer market have leveled out. With the rising popularity of imported beers, which typically come in glass bottles, domestic brewers have sought to capitalize on the prestige image of imported beers by returning to the glass bottle. Wine coolers, sold exclusively in glass bottles, are another relatively new market.

In addition, glass has benefitted from a number of new applications, particularly for microwave cooking. Kraft's Cheez Whiz product, which had been around for thirty-five years, enjoyed a 30 percent increase in sales after a new suggestion was printed on its label: "Try in microwave for instant cheese sauce." The product, sold in glass jars, illustrates the successful new marriage of glass and the microwave oven. Consumers have also indicated an overriding preference for cooked pastas and pasta sauces in glass containers, with 48 percent choosing glass over plastic (20 percent), metal (16 percent), and paper cartons (6 percent). Baby food also remains a product widely preferred in glass containers.

Perhaps surprisingly, glass has also seen a resurgence in the pet food market. Consumers are attracted to the glass containers because, unlike cans, they are resealable and leftovers can be stored in the refrigerator without creating an odor problem. In addition, of course, the glass container and its contents can be warmed in the microwave.

Glass is also 100 percent recyclable (although the recycling process is not without problems, as we shall see). Unlike paper, which is weakened by the process, it loses none of its strength in recycling and can be used over and over again to make new containers. Typically, glassmakers use 20–30 percent cullet in their glassmaking operations. However, industry leader Owens Illinois, which claims about a quarter of the glass container market, has produced glass containers made with 60 percent cullet, and has successfully tested containers made with 80 percent cullet. The cullet helps facilitate the melting of other raw materials and allows for furnace operation at lower temperatures. Every one percent increase in cullet saves one-quarter of one percent of the energy used to make glass containers. And it is estimated that using 50 percent cullet can double the furnace's life. Air emissions of nitrogen oxides and particulates are reduced, with total air pollution reduced as much as 20 percent by using recycled glass, and water use is cut down by 50 percent. In addition, one ton of cullet saves more than a ton of virgin raw materials.

Theoretically it is even possible to use 100 percent cullet, but there are a number of reasons why this is not practiced in the industry. The somewhat modest energy benefit of using cullet is more than offset by its higher cost; the raw materials for glass are inexpensive, approximately $40 per ton or about a third less than the cost of container-grade cullet. This higher price tag for cullet stems from the transportation and separation costs associated with glass recycling.

Another problem in glass recycling is that glass containers must be separated by color, and this (under present technology) has been invariably a manual process. Coloring agents are used to protect the bottled product from the effects of sunlight—green bottles, for example, contain ferrous sulfate or chromic oxide, while brown or amber bottles contain ferric oxide. However, these same beneficial coloring agents complicate the recycling process. Clear (or flint) glass containers can tolerate only minute quantities of brown cullet in the mix, less than 5 percent, and they can tolerate no more than 1 percent green cullet. On the other hand, flint cullet can be used in colored glass bottles by adjusting the coloring agents in the final mix; however, glassmakers prefer not to do so. Mixed cullet is generally avoided in glass container manufacture.

In addition to the necessary separation by color, flat glass must be separated from container glass. Flat glass has a somewhat different chemical formulation, particularly its higher calcia content, which affects the formability of the container glass. Recycling glass containers is a closed loop; bottles are made into bottles, flat glass trimmings go into flat glass. Both, however, could be used to some extent in fiberglass production.

Mixed cullet does have some uses, including construction and road-surfacing applications, although this lower-grade cullet has a lower market value: up to 50 percent less than container-quality cullet. Glass fiber insulation manufacturers can use high-quality flint cullet, though few are willing to use post-consumer cullet at all. Cullet has been used as an aggregate substitute in asphalt and concrete in Baltimore, Maryland, since 1971, and as much as 60 percent of some Baltimore streets are composed of glass. [1]

The easiest place to separate glass is at the household level or at the food service establishment, although collection agencies also do this work, by hand. Theoretically it is possible to separate the different colored glasses using light waves, since each color is associated with a different wavelength, and promising research has already been conducted in automated separation systems. None have found their way into commercial use, however, and hand sorting continues at the collection agencies.

Rarely today are glass bottles reused. The reusable bottle was the norm up until the postwar period; their subsequent decline was largely a matter of costs. Reusable bottles need to be thicker and stronger than one-way bottles. A thicker bottle takes more energy and raw material to make, and is heavier and costlier to ship. However, the potential overall energy saving of refilling versus recycling the bottle may help bring the refillable back into favor. It should also be noted that the glass melting process releases large amounts of carbon dioxide—and therefore poses the danger of contributing to global warming. The refillable bottle may hold out a solution to this problem.

Lightweighting is another important trend in the glass container industry, as it is in other materials sectors. The average 16-ounce nonrefillable bottle weighed just over 9 ounces in 1984; by 1987 the weight had been shaved to just over 7 ounces, and the downward trend is continuing. Glassmakers have learned how to maintain the strength of glass containers while making them ever thinner.

The surface condition of the glass largely determines the strength of the material. Tiny fissures or micro-cracks can occur as a result of abrasion or corrosion. The smooth surface of a new bottle, fresh from the mold, may withstand a pressure of 200,000 lbs. per square inch (psi); with handling, however, the surface becomes scratched and bruised, and by the time it reaches the store shelf, the strength of the bottle may be down to 3,000 psi.

Uniform distribution of the glass helps eliminate weak spots, or thin areas. Press-and-blow technology, whereby the gob is mechanically pressed against the empty mold and later blown against the final mold, has been found superior to blow-and-blow technology, in which the hot gob of glass is initially blown by compressed air over the empty mold. The press-and-blow method improves the distribution of glass, increases production speeds, and produces lighter and stronger bottles.

Coatings are also important in protecting the bottle. These are not designed to add strength to the bottle or combine chemically with the glass. Rather, they lubricate the glass surface so that contact with other bottles or with machines is less likely to damage the surface. Typical coatings are silicone, metallic oxide, wax, and sulfur compounds.

In the final analysis, perhaps the greatest competitive advantage that glass enjoys over other materials is its inertness to a variety of food and nonfood products. Acidic foods will readily react with metals and will alter the taste of the food. And components of plastics will leach into some foods, again altering the taste of the product. The clarity and visibility afforded by glass make it a premium packaging material for many foods and beverages. In addition, its impermeability to oxygen and other gases is important, especially for beer and carbonated beverages. It is a low-cost container, and its raw materials are in abundant supply throughout the world. To this can be added the fact that glass is fully recyclable.

On the negative side are the weight and breakability of glass. The expense of transporting cullet may make it less attractive for a glassmaker to use recycled materials, although it can be argued that the costs of not recycling are far greater to society. Glass consumes about 10 percent of U.S. landfills, and the country currently recycles only about 10 percent of its glass containers.

Economic and legislative incentives will be needed to spur glass recycling efforts, and further research is needed to remove some of the technical difficulties, such as separation by glass color. Recycling and also reusing glass containers would relieve the burden on landfills, reduce energy consumption, and lower water and air pollution.

Plastics

Plastics in their many forms are the third most common family of packaging materials, following paper and glass, as well as the fastest-growing. Plastics account for 14.5 percent of all packaging materials. In 1988, 54.5 billion pounds of plastics were produced in the United States, of which slightly over one quarter went into packaging. Of the 13.8 billion pounds that became plastic packaging, 7 billion pounds were used for containers, 5 billion pounds for films, 1.1 billion pounds for coatings, and 700 million pounds for closures.

There is a general lack of understanding of the complexity and variety of plastics. They represent an enormous variety of chemical compositions with an endless number of final applications, making them perhaps the richest and most complex of all packaging materials. As a consequence of this diversity, the full environmental consequences of the manufacture, use, and disposal of plastics are among the most challenging to grasp.

The plastics used in packaging may be one of several possible polymers. The following are the most common, in order of highest to lowest market share, along with examples of typical applications:

- *Low-density polyethylene (LDPE)*—a flexible packaging used in films, wraps, trash bags, and coated papers, as well as in some lid stock.

- *High-density polyethylene (HDPE)*—a translucent material used in milk and detergent bottles.
- *Polypropylene (PP)*—a stiff, heat- and chemical-resistant substance used in tubs and syrup bottles, as well as in films for food packaging.
- *Polystyrene (PS)*—a foam packaging material with excellent thermal properties, used in foam coffee cups, food trays, and "clamshell" hamburger boxes, as well as nonfoamed clear food trays and lids.
- *Polyethylene terephthalate (PET)*—a tough and shatterproof material used in soft drink bottles, as well as in food and medical containers.
- *Polyvinyl chloride (PVC)*—a clear and stiff packaging material used in water and cooking oil bottles as well as meat wrap.

The growth of the plastic container has to do with a number of appealing characteristics that have won it market acceptance: It is relatively unbreakable, and for this reason has displaced glass as the container of choice for shampoos, medicines, cleaning fluids, and many foods and beverages—anything that would shatter easily on bathroom and kitchen floors. It is lightweight and durable. It can be either clear or opaque, smooth or textured, white or brightly colored. It can be formed into any number of pleasing shapes. It is moisture-resistant, a major factor in the displacement of paper grocery bags by plastic and the widespread acceptance of plastic garbage bags.

Myths About Plastics

The public's negative perceptions about plastic are numerous. Perhaps the most common complaint is that it is nonbiodegradable. Florida State Senator George A. Kirkpatrick reportedly once asked, "For a hamburger that lasts a few minutes, why do we need a package that lasts as long as the pyramids?" [2] True, plastics do not degrade in landfills—but neither do hamburgers. As we have noted, virtually nothing breaks down in modern airtight and watertight landfills.

Because very little plastic is recycled in the United States, there is a widespread belief that plastics cannot be recycled. However, almost all of the plastics used in consumer packaging and disposables are thermoplastics, which means that they can be melted, remolded, and reused. Indeed, there is a value chain associated with plastic use and reuse (see Figure 3-3). Unfortu-

Figure 3-3. Value of different reuses of plastic.

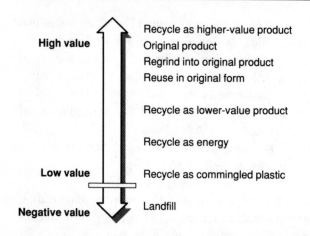

nately, regardless of their useful lifetime and recyclability, plastics today are most often destined to become a part of landfills, where they are second only to paper in volume. But, as the examples described later in this chapter show, many types of plastics are successfully being recycled into a variety of useful products. In fact, there is such active commerce in plastic recycling that current market prices for recycled plastics are regularly quoted in trade magazines.

Plastic is also widely believed to be a waste of our nonrenewable petroleum resources. However, the portion of U.S. petroleum and natural gas used to make petrochemicals is negligible, no more than 3 percent; less than 2 percent is used in plastics and only 0.5 percent is used in plastic packaging (see Figure 3-4). By contrast, almost half the nation's crude oil consumption is expended in the form of gasoline. Banning plastic packaging would have no appreciable effect on oil or gas consumption. An argument could also be made that society derives a greater, longer-term value from the small percentage of petroleum used in petrochemicals—and suffers far less pollution damage—than from fuel combustion.

Another recurring objection to plastics is that they are hazardous to manufacture and to incinerate. The process of manufacturing polystyrene foam has been associated with the release of chlorofluorocarbons, or CFCs, which have been identified as damaging to the earth's ozone layer. Recent

Figure 3-4. Uses of crude oil.

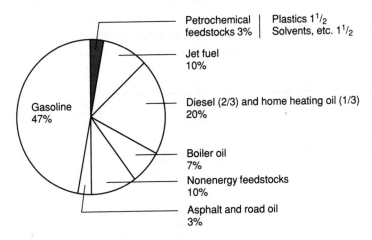

Petrochemical feedstocks 3% | Plastics $1^1/_2$ Solvents, etc. $1^1/_2$

Jet fuel 10%

Gasoline 47%

Diesel (2/3) and home heating oil (1/3) 20%

Boiler oil 7%

Nonenergy feedstocks 10%

Asphalt and road oil 3%

Less than 2% of crude oil is used for plastics.

Source: Mobil Chemical Company

advances in polystyrene foam manufacture have employed alternative blowing agents, however, and CFCs have been virtually eliminated from packaging applications.

Other plastics pose incineration dangers because they contain metal-based pigments or because, in the case of PVCs, they have the potential to release dioxins. However, technical advances such as scrubber systems and combustion controls that assure optimum temperatures are able to eliminate these concerns. (A fuller discussion of incineration appears in Chapter 4.)

Plastic Bottles

About 20 billion blown plastic bottles are produced annually in the United States. The HDPE bottle, used widely for milk and spring water, holds the leading market share by resin, with 80 percent of the plastic bottle market. The 2-liter PET bottle holds the second-largest share.

Plastic bottles are up to 90 percent lighter than glass, are shatter-resistant, and require less energy to produce than glass or aluminum. For example, PET containers are 25 percent more energy-efficient to produce than glass and 65 percent more efficient than aluminum. The plastics most often used in bottles, with some of their characteristics, are listed in Figure 3-5.

Figure 3-5. The properties of plastics used for bottles.

	Nitrile	Polycarbonate	PET	HDPE	Oriented polypropylene	PVC
Clarity	Clear	Clear	Clear	Hazy, translucent	Clear	Clear
Permeability to:						
- Water vapor	Moderate	High	Moderate	Very low	Very low	Moderate
- Oxygen	Very low	Moderate to high	Low	High	High	Low
- CO_2	Very low	Moderate to high	Low	High	Moderate to high	Low
Typical uses	Cosmetics	Mineral oil	Carbonated beverages	Detergents	Detergents	Cosmetics
	Household chemicals	Baby nurser bottles	Mouthwash	Bleaches	Drugs	Personal care
		Water	Liquor	Milk	Mouthwash	Household chemicals
		Milk	Edible oil	Chocolate syrup	Shampoo	Edible oils
			Drugs	Industrial cleaning powders	Household chemicals	Vinegar
			Cosmetics	Drugs	Intravenous solutions	
				Cosmetics	Liquid soaps	
				Lubricating oil		
				Edible oil		

There are limitations associated with plastic packaging. Plastic is a gas-permeable material, and high-pressure contents such as champagne would eventually lose pressure in plastic. By contrast, metal and glass are viewed as absolute or "functional" barriers, important in food applications in which oxygen transmission can affect flavor change. In the preservation of chemically active foods like pickles, some plastic packaging is not chemically stable. Conversely, lye is often sold in plastic bottles because metal cans cannot handle it for long-term storage.

Recycling programs currently accept many types of plastic bottles. Unlike glass bottles, which are converted into glass bottles all over again, plastic bottles can be transformed into a variety of recycled goods. In 1987,

72 million pounds of HDPE were recycled into furniture, pipes, signs, traffic cones, milk crates, toys, trash cans, and many other products (see Chapter 4 for more details on recycling).

PET, also known commonly as polyester, is a relatively high-volume, high-value plastic. It is second only to polyethylene in market share, although PET in both virgin and recycled forms is worth more than polyethylene. Besides the familiar 2-liter soft drink bottles, PET is used in biaxially oriented film (a modest-scale packaging application) and in fibers and fabrics (large-scale applications).

Recycling of the PVC bottle is far behind PET and HDPE, the two most recycled plastics. PVC is a more complex and difficult material to recycle, partly because of PVC's inherent thermal instability. In addition, a wide range of PVC compounds exist; because of plasticizers and other additives, there is no such thing as a standard PVC bottle.

PVC has also had more environmental pressure applied to it, especially in Europe. Germany and Italy have attempted to ban or eliminate key uses of PVC. This concern sprang from an incident at the Italian town of Seveso, where a large chemical spill released dioxin, a highly toxic, chlorinated compound, the same that is sometimes generated when PVC is incinerated. And since incineration is more widely used in Europe for waste disposal, the European public and activists have applied tremendous pressure on all chlorinated polymers, including PVC and PVDC (polyvinylidene chloride). In the United States, major producers like B. F. Goodrich, one of the largest PVC producers in the world, are working to establish a waste management infrastructure and technology for handling PVC products, including shampoo bottles as well as film products, blister packs, and other items.

The wide variety of plastic bottles has created some confusion in the minds of consumers, and this complicates the recycling effort. To help consumers sort out the alphabet soup of PVC, HDPE, PET, PP, and so forth, the Society of the Plastics Industry (SPI) has proposed uniform labeling guidelines for seven types of plastic containers (see Figure 3-6). The packaging industry has voluntarily put the SPI labels on the bottom of plastic containers, where they can be located readily, to aid in identification and sorting for recycling. The SPI is also working to expand the voluntary use of the code on flexible packaging.

Figure 3-6. Uniform SPI labeling guidelines for seven types of plastic containers.

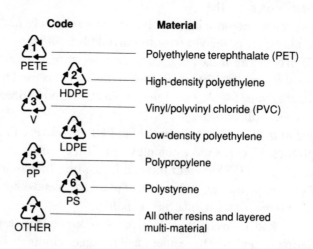

Code	Material
PETE	Polyethylene terephthalate (PET)
HDPE	High-density polyethylene
V	Vinyl/polyvinyl chloride (PVC)
LDPE	Low-density polyethylene
PP	Polypropylene
PS	Polystyrene
OTHER	All other resins and layered multi-material

One of the most significant plastic packaging innovations is the multilayer coextruded plastic bottle, which provides oxygen barrier properties previously unknown in plastics. This development will be described further under "Composites" later in this chapter.

Plastic Films

The earliest film products were made from organic materials. Cellophane is a wood-based material (the term itself is derived from the words *cellulose* and *diaphanous*). Most film products today are based on synthetic polymers and are often multiple layers of different plastics. Cellophane itself is often coated with polymers to improve its barrier qualities.

Film can be made by two basic methods: blown tubing or flat die. In the former, the resin pellets are melted and blown upward through a circular die into a tall vertical tube. Air pressure expands the tube while it is still hot. It can be oriented, or stretched, to strengthen the material. The final material can be rolled into tubing, or slit to make flat film. The flat die, or cast film, method also starts with melted resins, but the method more closely resembles papermaking; the material is cast onto a water-cooled roll or into a water bath to create a sheet of film.

Coextruded film is made by parallel dies that each extrude a layer of film, which may be of the same or different materials. The hot flowing resins are layered upon each other and fuse together, but they remain distinct layers. By contrast, laminates begin with separate layers of film to which an adhesive is applied in between the layers. This might be a cold, water-borne adhesive or a hot-melt adhesive system, which uses a layer of melted resin between the cooled layers.

Coextrusion offers a number of advantages over other multilayering methods. From a cost standpoint, only one step is required, and layers of inexpensive materials can be fused to produce a high-performance film.

The chemical compositions of plastic films are limitless and can be tailor-made to the most exacting specifications. Polyethylene is the largest volume plastic film produced in the United States. Low-density polyethylene might be used in high-clarity applications, such as bread wrap or soft goods packaging. A high-density polyethylene is desirable for food packaging that calls for high-moisture, gas, or grease barriers. Packaging films made from polyvinylidene chloride (PVDC), such as Saran®, are a well-known food wrap application with excellent heat resistance and barrier qualities. Other important types include nylon, polyester (PET), polycarbonate, polyvinyl chloride, polypropylene, and many other polymers and copolymers.

The impact of plastic films on the environment has received little attention, partly because they represent a smaller and less visible component of landfills. And for recycling purposes, their use in food and medical applications makes them challenging to clean and sort. The greatest stumbling block lies in the sheer difficulty of recycling anything flexible. Rigid materials can undergo granulations and separations. Flexible ones do not granulate easily and are not physically separable. The technology for recycling films in mixed plastic waste is only now being developed. As their use grows, and as their composition becomes ever more complex, plastic films represent a growing opportunity for recycling a material that has been an overlooked resource.

Issues in Plastics Recycling
Less than 2 percent of the plastic packaging produced in the United States

every year is recycled. Of that, PET is recycled more than anything else; 12 percent of recycled plastic, or 1 million tons, is PET.

The issue of finding food-contact applications for recycled plastic is an important one. The difficulty is guaranteeing a consistent reprocessing system in recycling, to ensure that no traces of problematic materials remain in the product. While recycled PET can be used in worry-free applications such as construction, autos, carpets, and lawn furniture, food is another matter.

In terms of energy content, plastics are petroleum-derived and have excellent fuel value in waste-to-energy facilities. And for mixed plastic waste, which is difficult to sort, incineration is a viable option and can reduce the volume of solid waste by about 75 percent. Polyethylene, polypropylene, and polystyrene have energy contents of 19,900, 19,850, and 17,700 British Thermal Units (Btu) per pound, respectively, while coal, newspaper, wood, and mixed solid waste have 9,600, 8,000, 6,700, and 4,500 Btu/pound, respectively. As indicated in Chapter 4, scrubbers, emission controls, and combustion controls are efficiently eliminating the hazards associated with incinerating plastic. The Wheelabrator plant in Millbury, Massachusetts, is running a 50-megawatt generator from mixed household waste. In all, there are eleven Wheelabrator plants in operation consuming 200 to 3,000 tons of trash per day and generating 5 to 75 megawatts of electricity. However, while all plastic materials do have a fuel value, greater value can usually be obtained by recycling.

As with any packaging material, numerous trade-offs must be made in plastic packaging: energy and water conservation versus minimizing the greenhouse effect; convenience versus recycling; recycling versus landfill reduction; food preservation versus costs. Which trade-offs are consumers willing to make? How many of us would be willing to revert to pre-1950s packaging, in which bread would go stale in a day, medicines were subject to tampering, and many fresh foods were simply not available?

Aluminum

Aluminum is noted for its light weight, good thermal conductivity, high reflectivity, resistance to oxidation, and superior barrier qualities. As such it finds a number of applications in packaging, ranging from beverage cans to aseptic containers. Beverage and food cans account for 85 percent of the aluminum used in containers and packaging. The remainder is largely used as household aluminum foil or as foil in flexible or semi-rigid aluminum trays, such as disposable pie tins (see Figure 3-7).

About 27 percent of all aluminum consumed in the United States is in packaging. In addition, most of the growth in aluminum consumption between 1979 and 1989 was in packaging, with an average annual real growth rate of 4 percent, compared to 1 percent in all other segments combined. Today in the United States, aluminum packaging is a 2-million-ton market, of which about 1.8 million tons goes into aluminum beverage and food cans.

Because of the importance of this market to the aluminum industry, it places a great deal of emphasis on continual improvement of aluminum as a package material (can, sheet, or foil). It has invested heavily to ensure that aluminum remains competitive, in its initial cost and the cost over its life cycle. Life cycle economics, combined with a genuine concern for the environment, led to the industry's development of national recycling strategies and networks. These have proven to be so successful that over 60 percent of the aluminum used in packaging is now recycled.

Figure 3-7. Uses of aluminum in packaging.

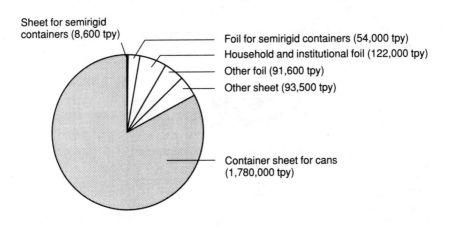

Sheet for semirigid containers (8,600 tpy)

Foil for semirigid containers (54,000 tpy)
Household and institutional foil (122,000 tpy)
Other foil (91,600 tpy)
Other sheet (93,500 tpy)

Container sheet for cans (1,780,000 tpy)

Source: Aluminum Association, 1989 Statistical Review

Because beverage and food cans dominate the use of aluminum as a package, the industry continues to devote most of its time and resources in these areas.

The Aluminum Can

While both Kaiser Aluminum and Reynolds were developing aluminum can technology in the late 1950s and early 1960s, it was Reynolds that produced the first production quantities of two-piece aluminum cans, in 1966. Prior to this time, cans were made primarily from tin-plated steel. Aluminum did not have the strength of steel per unit of weight, and could not be soldered. This led to the perfection of the two-piece drawn and ironed (D&I) method.

In all, nearly 112 billion cans were produced in the United States during 1989, and nearly 75 percent of these were consumed by the beer and soft drink industry. In terms of total units, aluminum dominated, accounting for nearly 75 percent of all metal containers.

In 1964, aluminum had held barely two percent of the beverage can market; today, 97 percent of beverage cans are made from aluminum. Aluminum won a lion's share of the beverage can market for a number of reasons. The two-piece aluminum cans were lighter and less expensive to produce than steel cans, and they were considered more attractive from a marketing standpoint (see Figure 3-8). At the same time, the aluminum can cut deeply into the glass beverage bottle market; the cylindrically shaped can is a low-cost package to produce, and is one of the easiest to decorate. It

Figure 3-8. Market shares of steel versus aluminum cans, 1990 (percent of total units).

	Steel	Aluminum
Beer	1 %	99%
Soft drinks	6	94
Food	92	8
General packaging	99	1

has a more cost-effective shape than the bottle in terms of volume, and is more efficient in terms of stacking. As of 1988, metal beverage cans, most of them by far being aluminum, far outdistanced glass beverage containers (see Figure 3-9).

Metal cans also play a dominant role in food packaging, but here, the all-aluminum food can has lagged behind; today only 8 percent of food cans are aluminum. However, the aluminum industry is targeting the food can for future growth.

A new can may be made from 100 percent used beverage cans (UBCs), and UBCs are the lowest-cost source of raw materials for aluminum production and subsequent rolling of can sheet. One of the reasons is that it takes 95 percent less energy to produce a pound of aluminum from UBCs than from virgin raw materials. Furthermore, the capital required for recycling is much less than that required for new primary (virgin) aluminum production. UBCs have been termed by some as the "aboveground mine."

The relatively high cost of the material has been the basis of one of the most important trends in aluminum package design: lightweighting. The idea is to produce a lighter can without affecting its quality. While it can be considered to be a form of source reduction, lightweighting was spurred by an economic incentive: to reduce material costs and thereby to encourage further use of aluminum cans. In 1980, 34 pounds of aluminum were needed

Figure 3-9. Package sales in 1988

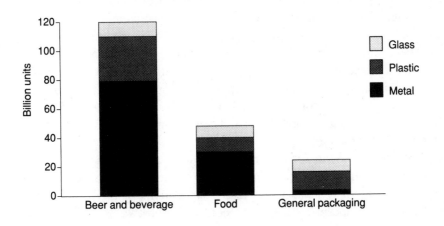

to produce a thousand can bodies. In 1988, that weight was down to 27 pounds, and by the mid-1990s it will be reduced to 25–26 pounds. While this kind of change has required billions of dollars of investment in manufacturing technology on the part of the aluminum industry and can manufacturers, it has ensured the continued dominance of the aluminum beverage can.

The top (lid) of the two-piece D&I can has also been engineered for lightweighting and material cost savings. Reynolds introduced the necked-in can in the late 1960s, and it was a breakthrough in technology and performance, especially in that it eliminated the cracked flanges that were common on straight-walled cans. In the 1980s, double-necked cans were introduced, then triple- and quadruple-necked cans, each successively reducing the diameter of the lid. Cost savings of 10–15 percent are driving the industry from the current 209/206 standard diameters to 204. Other innovations came in the can bottom; Alcoa reduced the bottom-heel radius while satisfying the buckle-strength requirement. The sidewall, too, has been redesigned. The so-called nominal thinwall, the area that has been thinned the most, has been reduced to a gauge of less than .005 inches. The result of all these innovations has been a significant reduction in material usage and costs.

Other Aluminum Packaging

Foil is another important aluminum packaging application. By industry definition, rolled aluminum becomes foil when it reaches a thickness of less than 0.006 inches. First commercially produced in the United States in 1913, aluminum foil was firmly established as a major packaging material during World War II, when it was used to protect food against vermin, moisture, and heat damage. After the war, the applications for aluminum foil boomed along with the economy. Semirigid containers first appeared on the market in 1948, initially as cost-cutting disposable pans for commercial bakeries; they continue in use for frozen foods and bakery goods. The food service industry began heavily promoting and distributing foil and semirigid containers the following year.

Packaging accounts for 75 percent of the applications for aluminum foil, with the remainder in decorative wrap, insulation, and other construction uses. Aluminum foil has found ready applications in the food industry because it is impermeable, greaseproof, nonabsorptive, inert, and highly formable, with excellent dead fold characteristics. For highly acidic foods, such as tomato sauce, aluminum requires the protection of a clear vinyl coating. Aluminum resists mildly acidic products better than mildly alkaline products, such as soap products. Concentrated mineral acids are not packaged in aluminum because of the possible corrosive effects.

Aluminum foil is also formed into caps, cap liners, and closures for beverages, yogurts, and other moist or liquid foods. Foil laminates and composites have been extremely important in tamper-resistant lids and form/fill/seal pouches.

Collapsible aluminum tubes used for toothpastes and medicinal products are being replaced by plastic laminates that are virtually unrecyclable, although aluminum tubes continue in use for most artists' paints. Tom's of Maine packages its "all natural" toothpaste in an aluminum tube to encourage recycling. Tubes in general seem to find wider uses in Europe, where they are used for tomato pastes, jams, jellies, and other food products.

Recycling

The incentives and economics of recycling are still evolving. Fundamental issues include the savings primary aluminum companies enjoy by using recycled material, the impact of state actions such as taxes or mandatory deposits, the local disposal costs, and the social benefits. The beverage industry, in particular promotes the recycling and recyclability of aluminum in order to avoid taxes or deposits and as support to the environment. The effect of industry participation has been nothing less than dramatic: The recycling rate has increased from 26 percent in 1979 to 61 percent in 1989. In 1988, the average recycling rate was 54 percent, although it varied by type of recycling program in place.

Although aluminum represents a small percentage of the municipal solid waste stream by weight—only 1–2 percent of total tonnage—it represents a large component by value. Aluminum is one of the most expensive and sought-after scrap materials, fetching forty-eight cents a pound in 1990, as against a mere four cents for steel.

Aluminum represents 2 percent of recoverable materials in no-deposit–law regions, but contributes an average of 33 percent of the material revenues for a curbside collection program (see Figure 3-10). Recently, several of the large waste management companies have signed agreements with the

Figure 3-10. Collection and processing stages in the recycling of used beverage cans (UBCs).

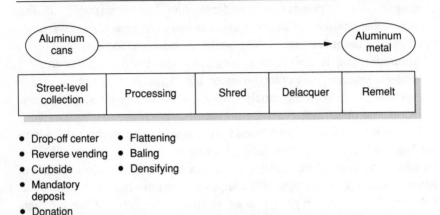

- Drop-off center
- Reverse vending
- Curbside
- Mandatory deposit
- Donation

- Flattening
- Baling
- Densifying

aluminum companies to turn over recycled cans. While all of the aluminum companies participate in recycling and collection, it is important to note that virtually all of the major bottlers and canmakers also are committed to this activity.

Recycling efforts in aluminum packaging have focused on the beverage can. Besides being the easiest and most visible to tackle, aluminum cans have the greatest potential impact on the environment; an aluminum can resists oxidation and degrades very slowly if exposed to the elements. Foil is difficult for the consumer to clean and separate, and it is harder still to extract it from the mixed waste stream. Recently, however, several of the aluminum companies have set up systems to deal with aluminum foil.

The environmental movement definitely has worked to aluminum's advantage. Indeed, it can be inferred that the reason why aluminum overtook steel so decisively in the beverage can market was its effectiveness at creating a recycling infrastructure. The first pilot aluminum can recycling center was established in Los Angeles in 1968, and recycling emerged as an important new source of post-consumer scrap. A record 49.4 billion aluminum cans were recycled in 1989, for a national recycling rate of 61 percent. However, there is no reason why the rate could not go much higher. Sweden, for example, has achieved a recycling rate for beverage cans of 75–80 percent.

For food and beverage applications, recycled aluminum is often used in direct contact with foodstuffs. Aluminum is melted at high temperatures, and anything organic and potentially harmful will be burned off. Thus there is no problem with recycled aluminum contacting food, even if the aluminum is uncoated, as is the case with aluminum foils and flexible trays. In addition, most aluminum cans, including beer and beverage cans, are coated with an organic resin to further protect the food product.

Outlook for Aluminum

In the future, aluminum will further erode the market for steel cans, especially in food applications. In 1990, aluminum held only about 10 percent of the metal packaging market for food applications, with the remainder dominated by steel. Steel has been the material of choice for food cans because of its low cost and strength. However, aluminum companies are developing cans that have the strength and retort to withstand the high temperatures and pressures of food canning procedures, including retort sterilization. Arthur D. Little calculates that aluminum will acquire 15 percent of the metal packaging market for food by 1995, while continuing to dominate the beer and soft drink segments. As illustrated by Figure 3-11, recycled aluminum is much more valuable in proportion to its weight than is glass, steel, or

Figure 3-11. Proportion of various materials recovered in curbside collection programs, compared with the revenues they bring.

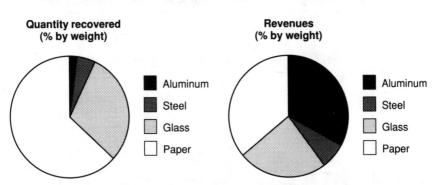

Aluminum represents only 2 percent of recoverable materials in a no-deposit-law region, yet contributes, on average, 33 percent of the revenues for a curbside collection program.

paper. Its package economics, recyclability, and high scrap value will continue to be important drivers in this industry.

Aluminum, to date, has not enjoyed similar successes in Europe or the Far East. Today, cans as a whole account for less than 20 percent of beverage packaging in Europe, compared to 70 percent in the United States. Only in Sweden does the can begin to match U.S. levels of penetration. Since 1983, however, cans have increased their share of the package mix by 50 percent (see Figures 3-12 and 3-13). The nondetachable top, a feature which was

Figure 3-12. Percentage of soft drink sales in Europe in glass, cans, and PET, 1983 vs. 1989

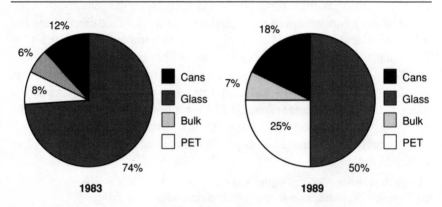

Figure 3-13. Percentage of beer sold in cans, selected regions, 1980–1989.

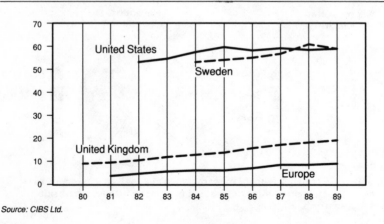

Source: CIBS Ltd.

standardized in the United States long ago, has only recently enjoyed a wide application in Europe.

Steel

A Parisian chef named Nicholas Appert devised the first canning process in 1809, and a year later Peter Durand, an English inventor, created the "tin canister." This technology was brought to Boston in 1817 by William Underwood, and the U.S. canning industry was born. Steel accounts for 6.5 percent of all U.S. packaging materials, compared with 2.3 percent for aluminum, although it has lost considerable market share to aluminum. As of 1989, the United States devoted 5 million tons per year of steel to containers, packaging, and shipping materials. Of this, an estimated 50 percent was for tin-plated steel cans.

Steel is recognized as a tough, long-lasting packaging material, impervious to air and water, making it ideal for long-term food storage. Steel cans for foods and juices are still mostly a three-piece construction (see aluminum cans), but the technology has improved to the point that two-piece cans can also be made from steel. These two-piece cans are used for premium applications like beer and other beverages and are almost indistinguishable from aluminum cans. Ends in paper cans and drum lids are also major outlets for steel. In these cases the side of the can or drum is made of spiral-wound paperboard resulting in a composite system with special recycling and reuse requirements. Composite structures are discussed in detail later in this chapter.

Their formability, high specific strength and stiffness, and highly developed seaming technology has lead to an extremely large variety of shapes and sizes of steel containers, ranging from the common 55-gallon drum down to "pill boxes" capable of holding less than an ounce. Specific examples of this versatility of function include:

Shape	Application
Square breasted	Baby powder
Oblong, hinged	Bandages
Oblong, F-style	Paint thinners
Three-piece sanitary	Foods

Two-piece sanitary	Foods
Aerosol can	Hair sprays
Flat, hinged-lid	Aspirin
Flat, round	Shoe polish
Oval	Lubricating oils
Pear-shaped	Hams

The largest application of steel packaging is for food, including vegetables, fruits and fruit juices, soups, coffee, dairy products, meats, poultry, fish, shortening, spices, nuts, and pet foods. The market for steel beverage cans is shrinking, although there are a few holdouts. Pepsico uses steel in about 10-12 percent of its cans. And a few national beer brewers use steel, such as G. Heileman Brewing Company for its Iron City label.

Other important steel packaging applications include motor oil, paint products, aerosols, waxes, household oils, powders, and lighter fluid, although plastics have been making inroads into some of these markets. Steel drums are an important, $200-million-plus market and have historically been one of the quiet success stories in recycling. In recent years, however, they have been associated with hazardous wastes, and drum reconditioners have become more and more reluctant to accept them. Their recycling has always been a highly specialized business.

Most food is packaged in tinplate, for its resistance to corrosion. So-called tin cans are 99.5 percent steel, coated with a very thin layer of tin. The relatively high cost of tin has led to a search for alternative coatings, to create what is called tin-free steel, or TFS. Perhaps the most promising is chromium oxide. First used in Japan and subsequently improved in the United States, a chromium coating is treated to produce chromium oxide on the can's surface. The result is a coating one-hundredth the thickness of tinplate, or about twenty atoms in thickness.

In addition, the interiors of cans are often coated with either organic or inorganic materials to protect and seal the contents inside. The wide variety of products packed in tinplate gives rise to an endless number of variables that influence corrosion. Highly alkaline products, usually nonfoods, will strip tinplate and organic coatings, but not the base steel. Highly acidic products are corrosive and may create a sulfide black condition, which affects both appearance and flavor. Highly colored acid foods, such as cherries or strawberries, are susceptible to color changes, and organic coatings were first used to protect red fruit colors. Very complex organic coatings are needed to protect soft drinks, which are corrosive, and beers. Beer, though not corrosive, is sensitive to steel, and off flavors can be detected even with very low iron contamination levels.

Steel materials of all types are recycled at a high rate—old cars, construction materials, and also factory scrap from steel plants. Scrap is a very important source of raw material to make new steel, and the United States is, in fact, a major exporter of scrap iron and steel. For recycling purposes, collection and sorting become a challenge if steel cans are mixed in with standard solid waste. Voluntary segregation is helping to improve the rate of steel can recycling. However, the deposit laws that have accelerated the recycling of glass and aluminum generally do not apply to steel cans. In addition, steel cans, containing the remains of everything from dog food to creamed corn, are not as easy or as pleasant to clean as beverage cans.

In order for steel cans to be reclaimed, they must be segregated from raw trash, ideally in curbside collection programs. Steel has the advantage of being easy to separate from other municipal waste with a magnet. If relegated to waste-to-energy facilities, steel cans can be extracted after incineration from the ash. However, incineration deteriorates the quality of the material, rendering it of little value for reuse.

The biggest disadvantage of steel recycling is that the value of scrap is low in absolute terms, compared with transportation costs, and relatively low compared with the value of new product. Scrap is not needed to make new cans, and other sources like construction debris and automobiles are readily available.

In the reprocessing of the material, the cans are melted at very high temperatures that burn off or vaporize the organic material. Some of the tin will also be vaporized, and some will alloy.

All in all, the economics of steel recycling have not been encouraging. The costs of collecting steel cans, cleaning them, removing labels, and shredding them for reuse are harder to justify when the base material costs a few pennies a pound. Nevertheless nearly 5 billion steel cans were recycled in 1989. With the advent of comprehensive recycling laws or systems that incorporate all types of packaging, including steel, needed infrastructures can be put in place to substantially increase the recovery rate of the lowly tin can from the municipal waste stream.

Composites

Few materials by themselves meet all the possible performance requirements of packagers at a reasonable cost. These requirements might include water vapor barriers, gas barriers, strength, flexibility, rigidity, clarity, opacity, inertness, or any combination of such criteria. But what no single material could accomplish alone is now being accomplished by composite materials—combinations of different packaging materials as multiple layers—which have been developed to meet virtually any packaging specification.

Composite materials are controversial from an environmental standpoint. For example, multilayer coextruded plastics or multilayer juice boxes are difficult to separate into their component materials for recycling. As a result, they typically end up in our already overflowing landfills. While some recycling applications have been developed from mixed materials, such as plastic lumber, these have often been low-grade end products with limited applications. A far greater value could be extracted from the separated component materials.

Composites are the most dynamic segment of the packaging industry. Materials are constantly changing in terms of composition, thickness, laminations, or coextrusions. The following are some of the most important categories of composite packaging materials:

High-Barrier Plastics

High-barrier films and containers are an important and fast-growing application of coextruded plastics. Estimated at 132 million pounds in 1990, plastic barrier packaging is expected to grow to 187 million pounds by 1995, a 7.2 percent annual increase, or more than twice the growth rate for plastics in general.

High-barrier plastic bottles consist of several layers coextruded into one structure. Each layer is intended to serve a particular function. A polypropylene layer would provide a moisture barrier; a layer of EVOH (ethylene-vinyl alcohol, a hydrolyzed copolymer of ethylene and vinyl acetate) would afford an oxygen barrier. A typical six-layer structure might have an outer layer of polypropylene, a recycled material layer, an adhesive layer, a barrier layer of EVOH, another adhesive layer, and an inside layer of polypropylene.

In 1983, the squeezable ketchup bottle was among the first users of multilayer barrier containers. Today, sauces, salad dressings, and jellies are on the growing list of barrier container applications.

High-barrier films have been used in the United States for specialty vacuum-packed meats, such as tenderloin, smoked turkey, cured meats, and premium cuts, as well as for fresh and smoked fish. These tough, clear, multilayer films can extend the shelf life of fresh meats to several weeks; in conventional supermarket wrap, the same meats might last a matter of days.

Some high-barrier packages have two kinds of multilayer film: a clear top layer and a printed bottom layer. Each construction has two or more component materials for which many constructions are possible—for example, PE/PVDC/PE. The fastest-growing barrier films, nylon and EVOH are expected to grow 15–20 percent from 1990 to 1995.

These packages are used increasingly in Scandinavia for fresh fish, as in other parts of Europe where refrigerator shipping is still limited. They are associated with perishable foods of a high nutritional value, and they often carry strong ethnic attachments—Germans and sausages, Scandinavians and fish. However, they also present interesting trade-offs. On the one hand, high-barrier packages preserve wholesome products, cut spoilage, and reduce the necessity for daily shopping for fresh meat and fish; in addition, they improve the barrier characteristics of the package, so that meat packers are able to rely less on nitrites and chemical preservatives. On the other hand, these packages pose the same difficulty as all composite materials; the components are difficult to separate and are currently unrecyclable.

This raises interesting questions for consumers and packagers: Should we accept a package that is less environmentally friendly, if it means we can improve a fresh food supply and eliminate potentially dangerous chemicals from our food?

Aseptic Packaging

The single-serving drink box is the best known application of aseptic pack-aging. The airtight box can preserve its contents for months without refriger-ation, and it is widely used for juice products in the United States. The container is widely used overseas for other products as well, including milk, wine, edible oils, coffee and tea, sauces, and soups.

Aseptic processing refers to ultra-high heating of a product and then immediate cooling. The aseptically sterilized product is combined in a ster-ile environment with a multilayer presterilized container. In the drink box, beverage is in direct contact with a polymer layer; there is also a layer of aluminum foil or other barrier material, paperboard, and another polymer layer. Aseptic packages may also be all-plastic, coextruded, thermoformed containers.

One such package is, as we have mentioned, the six-layer "Tetra Pak" developed by Swedish inventor Ruben Rausing in the early 1950s. The box's outermost layer is polyethylene, followed by single layers of paper, polyethylene, and aluminum foil and then two layers of polyethylene. In 1990, Tetra Pak Rausing, which owns and licenses the technology, reported revenues of $5 billion from sales of 60 billion cartons in 109 countries; it also claims royalties from boxes not produced by Tetra-Pak.

While the Tetra Pak is by no means the only multimaterial package, its sheer popularity has made it a focus of environmental debate. The state of Maine banned the box in 1990 because environmentalists cited it as nonre-cyclable, and other states were considering similar measures. U.S. drink box makers bounced back with an aggressive advertising campaign, making the case that the boxes are recyclable into such products as artificial lumber. The industry also subsidized pilot recycling projects for the boxes in schools. However, the industry's claims met with a complaint filed by the New York City Commissioner of Consumer Affairs, charging deceptive trade practic-es. Environmentalists have insisted that recyclability claims are misleading if the consumer has no local opportunities to recycle a product.

It should be noted that, while the useful market for artificial lumber made from aseptic boxes is limited, other materials, including aluminum and paper pulp, can be extracted from the aseptic box, and could yield a far greater value. The drink box remains an environmental "hot button" that will be a harbinger of things to come for the packaging industry.

Some other multimaterial packages include:

- *Composite cans.* Spirally wound paper containers with one or both end closures permanently affixed. Refrigerated pastry dough and frozen orange juice are two of the most familiar applications. While paper is the primary component, the body of the container is based on several layers including foil and plastic. The end closures may be metal, plastic, paper, or combinations of the above.

- *Form-fill-seal pouches.* Flexible, multilayer pouches used increasingly in single-serving food products, as well as in premeasured pharmaceuticals, including medications and medical supplies such as sutures, clamps, and other devices. These pouches are designed to provide protection from physical damage as well as from bacterial contamination. The choice of materials depends on the product. For example, a sterilized food might require a heat-sealable layer such as polyethyelene. A simple example might be a paper/LDPE/aluminum foil/LDPE composite, for moisture-sensitive dried foods.

- *Carded packages.* Paper/plastic packages designed to display a product in a clear plastic material, mounted on a paperboard surface that contains printed information. There are two types: Blister packages use a rigid thermoformed plastic blister that is heat-sealed to the paperboard; skin packaging involves draping heated film over the product and its mounting, which draws down tightly to the board and the product. Polyethylene and ionic polymers, such as Surlyn®, are two of the most common skin packaging films, while blister packs may use one of a number of styrenes, cellulosics, or vinyls.

- *Ovenable trays.* Disposable ovenable trays designed to withstand microwave or conventional oven cooking. These may be a paperboard construction with an extruded PET coating; they might also be made from solid PET, a reusable product.

Closures

Screw caps, lug caps, crowns, and other closures have typically been made from metal or plastic, although today they are often made with both. Metal caps can be supplied with plastisol liners to allow a hermetic seal. Injection-molded plastic caps may be used without liners. Innerseals of foil, paper, plastic foams, or other materials provide additional barrier qualities, as well as a "safety seal." The rise of the tamper-resistant or tamper-evident bottle has meant the addition of more materials to package closures. While a small part of the waste stream, closures and their increasingly complex components raise additional environmental challenges.

Laminations and Coatings

Lamination is an alternative to coextrusion in combining the properties of various materials in flexible packaging. Many products demand special barrier properties; the most common elements for which a barrier is required are water vapor, oxygen, odor, flavorants, and light. Other desired performance characteristics are strength and heat sealability. Laminates help meet these requirements at a reasonable cost.

When aluminum foil is rolled to thinner gauges, small holes may occur in the material. Lamination of plastics to the foil can alleviate the effect of these pinhole defects. And when paper is laminated to foil, for example, the resulting product has the stiffness of paper and the barrier and deadfolding qualities of foil. Margarine wrappers, for example, are a lamination of foil to paper tissue.

Laminates are created by one of four methods: wet bonding, dry bonding, thermal laminating, and hot-melt laminating. The first two involve the use of adhesives. Thermal lamination seals a film to a substrate by heat. Hot-melt lamination involves the application of a melted layer.

Coatings are a specialty application. They may be applied to protect printed surfaces and provide water resistance; wax or polyethylene are the most typical. Or coatings may be a base for printing. A clay coating may be used on paperboard applications such as cereal boxes. Ultraviolet (UV) coating, which requires ultraviolet light to cure and set, is used on cosmetics boxes and other upscale packaging applications, which favor the crisp, highly glossy graphics the coating affords. These coatings are to be considered separately from laminations.

Metalizing is a new coating technology that involves spraying a thin layer of aluminum on paper or plastic films. The metal layer is extremely thin, about 300 times thinner than the thinnest commercial foils. The metal is melted, vaporized, and condensed on the substrate. In addition, an outer layer of reverse-printed, non-heat-sealed films will be laminated to the metalized surface. The result is a good barrier against moisture, gases, and light. Fatty and fried foods, such as potato chips and nuts, as well as coffees are protected from oxidative rancidity.

Biaxially oriented nylon (BON) or oriented PET (OPET) are the metalized substrates used for coffee containers; metalized LDPE is used for hosiery packages; chips, nuts, and candy are packaged in metalized OPET or biaxially oriented polypropylene (BOPP). These packages have the shine of metal but are tougher than foil, and they have the barrier qualities needed to compete with foil.

The Japanese have been active in silicone deposition technology, using the same principle as metalizing. Silicone is vaporized and deposited on a substrate, providing the barrier and clarity of glass in a flexible material.

While coatings and laminates provide performance characteristics previously unknown, their environmental consequences are also uncertain. Paper recyclers have devised methods to remove polyethylene and wax coatings, but more complex materials such as metalized structures currently defy commercial recycling efforts.

Labeling

Labels provide information, identification, and graphics to packaging, and they usually involve an additional material as well. They may be paper, foil, or plastic applied with hot-melt or cold set adhesives or with heat-activated sealants.

Sleeve labels have found increasing usage for pharmaceuticals and food products that are vulnerable to tampering. Preprinted polyethylene, PVC, or vinyl copolymers are fitted around the entire package, and then they are heat-shrunk.

Prelabeled plastic, in which the label is molded into the plastic layer (applied to the container as it is blow-molded) saves a separate labeling step and replaces a small amount of the container plastic. These can be shipped already loaded in cartons for in-case filling.

Avery Dennison Company has produced a number of environmentally sound labeling concepts, such as a 100 percent polyethylene label that can be recycled along with the polyethylene container. The pressure-sensitive polyethylene label, called Fleximage, was also designed to be conformable to irregularly shaped plastic containers. It was invented by William Ewing and is licensed to Avery Dennison.

Closing Thoughts

The packaging industry will need to take a great many more criteria into consideration in package material selection and design. Composite materials pose the greatest challenges for productive reuse of materials, although opportunities exist for innovative solutions. Perhaps a better alternative is to combine materials that can be more readily disassembled, or that are more compatible with each other for recycling purposes.

It may be argued that there are environmental benefits associated with such packages as the aseptic juice boxes: dramatically lower in weight and volume than glass and metal containers, they take up less space in landfills; they also cost less to ship, and the fuel energy expended on shipping cannot be recovered. At the same time, they bring cost savings, efficiencies, and convenience benefits to consumers and suppliers. In the long run, however, packaging companies will need to devise ways to separate composite materials, to meet the public's more exacting environmental demands—and to stay competitive.

In addition, the materials-handling technologies for post-consumer plastics are primarily geared to rigid materials, such as bottles and cans, rather than to flexible packaging. There is need for recycling technology in the flexible packaging area. Recycling innovations are especially needed in the complex high-barrier segment, which combines the difficulties of multi-component materials and films.

Technology transfer has been important in the development of composites. Japan has been the source of much research on new unique materials. Europe has also been the source of major innovations in composite materials. To find viable solutions for post-consumer uses of complex materials, we may also need to look globally for sources of innovation.

Notes

1. Susan Selke, *Packaging and the Environment* (Lancaster, Pa.: Technomic Publishing, 1990), p. 108.
2. "Plastics and the Environment," Society of the Plastics Industry, 1990, p. 30.

4

Infrastructure: Limitations and Opportunities

One of the major challenges society faces is building an infrastructure for sound environmental practices. Without it, all of the environmental consciousness of corporations, legislatures, and consumers amounts to no more than good intentions. For businesses, it means finding ways of making a return on investment while making thoughtful use, and reuse, of materials so that the recycled container and its by-products pay their own way back through the system. For government, it means creating the incentive systems to support and encourage such an infrastructure and to distribute responsibility fairly among all participants.

Modern industrial society has excellent distribution channels to get products to their destination, either to retail stores or to consumers. And, of necessity, distribution systems have been created to move trash from consumers and stores to landfills and other disposal facilities. However, there is currently little incentive in place to drive the next and final link in the chain: getting usable material from the collection site back into the system for recycling or reuse.

Government has supported the development of railroads, highways, and overland transport systems through subsidies and taxes, and this in turn has fueled industrial growth. In a sense, traditional U.S. incentive systems have been one-way, and product-pricing structure has included the cost of transportation. Can the government now apply incentives and innovation to create a kind of reverse distribution and collection? After all, if it is possible to distribute a product over a wide area to a diverse population of end users, why should it not be possible to get the refuse from that product back to its source?

On closer examination, the problem is much more complex. Recycling is not always a closed loop; that is, PET bottles are generally not remade into new PET bottles (although aluminum cans are typically recycled into more cans). Is recycling less meaningful when plastic milk jugs are converted into detergent bottles and others into flowerpots and building materials, and virgin material is required to make each new milk jug?

This is but one of the issues to be faced as our society works to build a sound waste management infrastructure. A partial infrastructure exists for reclaiming certain packaging materials, as well as for incineration and other disposal options. We will explore this existing infrastructure and the challenges it poses.

Waste Management Options

In spite of many isolated success stories, we are still very far from achieving an integrated, comprehensive system of waste management on a national level. The U.S. Environmental Protection Agency has defined a hierarchy of integrated waste management systems, with an emphasis on the first two:

1. Source reduction, including reuse of packaging
2. Recycling, including composting
3. Waste combustion, with energy recovery
4. Landfilling

At present, the United States depends most on the fourth solution, relegating 73 percent of its trash to landfills while incinerating 14 percent and recycling 13 percent. While recognizing that the bulk of waste will continue to be managed through landfills and combustion, the EPA has recommended a significant shift of the nation's municipal solid waste to source reduction and recycling.

The most effective means of solving the garbage crisis is to prevent it at the source. In packaging, source reduction may be achieved at the corporate or consumer level through selective purchasing and through the reuse of packaging materials. Lightweighting is an important form of source reduction, although the motivation behind it has traditionally been economic and not environmental. In addition to minimizing waste, lightweighting lowers material costs and reduces shipping costs for producers and distributors.

At the same time, however, the hectic life-style of the two-income family has fueled the explosive growth of microwavable dinners, single-portion lunch packs, and other convenience packaging. Other products, such as toys, are packaged in oversized containers designed to grab the attention of shoppers, dominate the store shelf, discourage shoplifting, and display advertising for the product. This trend has coincided with the rise of the toy "supermarket," in which hundreds of products compete for shelf space. Hardware stores are another kind of specialty supermarket, in which wooden bins of nails have been replaced by blister-packs of nails in small quantities. Without a dramatic change in the consumer's desire for convenience and in the way products are displayed and sold at the retail level, the excessive package will continue to be favored by both marketers and consumers.

Effective source reduction slows the depletion of resources and prolongs the useful life of the available waste management infrastructure.

Recycling

In order for recycling to succeed in today's world, there must be a consistent and reliable source of supply, suitable processing facilities to convert the recovered materials, and a ready market for the recycled product. Failing any one of these components, a recycling system cannot be sustained.

Effective collection systems are the key to providing a reliable supply of materials. In mandatory deposit law (MDL) states, the five- or ten-cent deposit per container is a powerful incentive for recycling, resulting in return rates as high as 90 percent. In non-MDL states, mostly in the western region of the country, the Beverage Industry Recycling Program (BIRP) has been established, with beverage industry assistance, to provide drop-off recycling centers for a variety of materials. These centers pay consumers for returned materials, although the payment is generally less than the typical five-cents-per-container in MDL states. As an additional incentive, BIRP centers are often combined with amusement parks and raffle events to encourage family participation.

Curbside pickup programs are even more successful than drop-off facilities, providing the most convenience to the consumer. In addition, the colorful and highly visible recycling baskets at curbside proclaim who in the neighborhood is recycling and who is not, adding a healthy measure of peer pressure to the process.

The economics of recycling are potentially more favorable than either landfilling or incineration, although they vary depending on the region of the country. In New Jersey, the cost of curbside pickup of recyclables was estimated at $50 per ton, after deducting the revenues from sales of recovered materials, compared with a cost of nearly $100 per ton for landfill disposal. In Rhode Island, the net savings of curbside recycling versus landfilling was put even higher, an estimated $56 per ton. Another estimate puts the cost of weekly curbside collection at $20-30 per ton, landfilling at $40-60 per ton, and incineration can be as high as $90-110 per ton. [1]

In addition to the "carrot" of cost savings, some communities are providing additional incentives, or "sticks," for recycling: charging consumers by the pound for nonrecyclable refuse, while sorted recyclables are picked up for free. Other communities, like New York City, are levying fines for the discarding of recyclable material. The greatest value of these programs is educational: They associate the generating of unnecessary garbage with a direct cost and consequence.

The wide fluctuations in the value of recycled material can wreak havoc on the economics of recycling. Both supply and demand need to be stabilized to avoid gluts and shortages and to minimize price volatility. To stimulate the market for secondary materials, the EPA has favored such government actions as procurement of recycled products, low-interest loans for the construction of recycling facilities, and tax incentives for recycling industries that relocate to communities where they are needed. These ideas coincide with private-sector moves to create a demand for recycled materials by making a commitment to packaging with recycled content.

In addition, the EPA has favored waste exchanges, whereby businesses can divert reusable materials to other businesses for reuse, rather than relegating them to the garbage heap. Candidate materials include paper (office stationery, computer printouts, newspaper, corrugated board), glass, plastic, compost, aluminum, steel, oil, and tires. The principle of "one man's garbage is another man's gold" is practiced by companies like Du Pont, which has found buyers for materials it once discarded. There are also opportunities for small businesses, like Flores Paper Recycling, Inc., of Miami, which sources office paper by combining services like document shredding with paper collection and by helping companies establish paper recycling programs.

Recycling can divert potentially large volumes of waste from landfills and incinerators, conserve natural resources, and serve as an educational tool to raise public awareness of waste management issues. We will now look to the recycling infrastructure on a material-by-material basis.

Metals

Ever since the days of the scrap collector in the horse-drawn wagon, there have been iron and steel salvaging operations in every major municipal area. Steel drums, for example, have been successfully refurbished and reused since the turn of the century, although the original impetus was not environmental consciousness but a practical business opportunity.

About two-thirds of all steel products in the United States are recycled every year. However, most of this recycling is outside of the packaging world: automobiles, industrial equipment, old rails, and steel structures are part of the steel recycling mix. Only one in five steel cans is recycled.

Steel scrap processors are formidable operations with extensive capital investments in machinery. Automobile hulks are fed into machines called automotive shredders, which reduce both ferrous and nonferrous materials to a manageable size before shipping them to steel plants and other recyclers. These scrap processors are ubiquitous and handle all kinds of steel products: construction materials, steel doors, radiators, and kitchen sinks. So why are so few steel cans being recycled? Part of the problem has to do with the fact that scrap processors are designed to accrue value from high-density, large-volume materials.

There is also the matter of public perceptions. Steel does not typically spring to mind when the consumer thinks of recyclable packaging materials. In addition, the household that willingly rinses out its beverage cans and bottles for recycling may be less willing to clean out dog food cans and other food containers—the domain of steel packaging.

To combat this second problem, the steel industry has launched a major effort to recast the image of steel packaging as "environmentally friendly." In 1988, the Steel Can Recycling Institute was formed, with steel industry backing, to spread the word about "America's most recycled material" in public awareness and advertising campaigns.

For the steel industry, the danger is that of losing market share to aluminum and other packaging materials. However, even if the general public comes to regard steel as an environmentally friendly food container, there are other factors that influence market share. State and local legislation may impose deposits on steel containers if they do not meet minimum recycling rates. In 1990, New Jersey set a mandatory recycling target of 15 percent, to be increased to 25 percent within the following two years; at the same time, Florida aimed for a 30 percent reduction in municipal solid waste through recycling, with strict penalties on industry sectors that do not meet the recycling targets. In 1989, California attached a two-cent deposit, or "recycling fee," to any type of container that did not meet a 65 percent recycling rate, and the rate would be raised to three cents if the target was not met by 1992. The fee would be reimbursed when the container was returned for recycling, with unclaimed fees going to support recycling programs and litter cleanup. The result of such laws is an economic disincentive to purchase steel cans if they fail to meet recycling targets.

The steel industry has undergone a considerable reversal of fortune since the 1960s, when virtually anything that was potable came in a can. While steel remains the metal of choice for food containers, the aluminum industry, as we have noted, has taken over 97 percent of the beverage can market. Aluminum suppliers accomplished this prodigious feat in part by aggressive advertising followed by building a recycling infrastructure, precisely the area that the steel industry neglected.

The key to aluminum's success was the establishment of a widespread collection network of 10,000 can buy-back centers. As an incentive to collection efforts, aluminum makers were able to pay a high premium for recycled cans, and yet benefit financially. Throughout the United States, including states without deposit laws, nonprofit groups from Boy Scouts to churches, along with individual citizens, collected $900 million worth of aluminum scrap in aluminum can collection drives in 1989.

Aluminum scrap is a valuable commodity because using recycled aluminum consumes substantially less energy and is significantly less capital intensive than processing virgin metal. Aluminum companies therefore have an incentive to invest in the collection and transportation infrastructure to ensure their supply of valuable scrap metal. In 1990, recycled aluminum cans fetched forty-eight cents a pound, twelve times the going rate for steel cans.

To retain its share of the packaging market, the steel industry is working to encourage food and beverage companies to use steel packaging and urging steel mills to accept materials derived from municipal solid waste. For example:

• Weirton Steel Corp., a major supplier of tin-coated steel, boosted the price it pays for used steel cans by 70 percent, to 7.5 cents per pound, at a time when the going rate for steel cans was four cents a pound. Its goal is to make steel more competitive with aluminum and to help create a market for collecting tin cans.

• USX Corp., the nation's largest steel company, installed 125 vending machines in the Pittsburgh area to collect steel cans for reprocessing.

• Bev-Pak, Inc. of Monticello, Indiana, established mobile steel scrap collectors and launched more than 800 steel can buy-back centers in the Midwest. The company makes more than 1.5 billion steel beverage cans a year, and it has found a market niche for steel beer cans in the Midwest.

Meanwhile, aluminum recycling does have an important drawback: Several different collection systems are in place, and as yet they are uncoordinated. First there is the system set up by the aluminum companies to collect and pay for returned cans, especially in states where there are no deposit laws. Second, in the so-called bottle-bill states, metal cans are collected and delivered to central collection points along with plastic bottles; it is usually not the aluminum companies that are in this collection business. The traditional scrap dealer purchases these materials and consolidates them for shipment. The latest approach in municipal solid waste management is to require consumers to segregate their trash, even in states without bottle bills, so that collection is done at the curbside; this creates the possibility for yet another system to collect aluminum.

Paper

Like aluminum cans, paper has long been the beneficiary of informal community recycling efforts, such as the familiar "newspaper drives." In addition, the containerboard business has formal and well-established mechanisms for collecting corrugated boxes from commercial and retail establishments and delivering them to collection sites.

The successful recycling of brown boxes (containerboard) has much to do with the nature of the material. Brown boxes are easy to identify and collect from convenient collection points, such as shopping malls, and this has spawned a large number of collectors. U.S. containerboard is also a high-quality material, making it an attractive export to other countries that recycle the material for their own use. The U.S. exports over 6 million tons of waste paper a year; from the port of New York alone, 1 in every 4 tons of export cargo is waste paper. Thus, by virtue of the quality of the material, the United States recycling infrastructure for containerboard extends to other countries.

Containerboard companies are changing the way they make the product and adding more recycled material to accommodate environmental concerns. Some, like Jefferson-Smurfit and Stone Container, are establishing collection systems themselves to control the supply of used packaging material. It is in their interest to do so, because of the tremendous price volatility in waste paper. Jefferson-Smurfit also has the distinction of being the largest owner of paperboard mills for recycling paper.

For paper other than corrugated board, recycling is limited by insufficient reprocessing facilities. Newspapers are easy to identify and collect from households and are typically recycled into chipboard or into new newsprint. However, de-inking is required to make new newsprint, and few operations have yet to invest in the necessary de-inking equipment—potentially an $8 million capital investment for an average size plant.

In an already capital-intensive industry, most paper companies are reluctant to make large expenditures for equipment that may grow quickly obsolete because of unforeseeable market changes. Uncertainties about demand for recycled paper products, as well as the price volatility in recycled paper, contribute to this reluctance. Indeed, some of the players in this market are unconvinced that the environmental movement is here to stay.

Beyond corrugated board and newsprint, the infrastructure for recycling other types of post-consumer paper has yet to be built. Nevertheless, the industry may have little choice in the matter. The public is demanding more and more that consumer product companies make use of recycled paper products and give something beyond lip service to environmental concerns. The companies are, in turn, exerting pressure on the paper mills. These market pressures are being matched by legislative pressures.

Scott Paper, which is a paper company as well as a consumer products company, is in a position to understand these pressures well. Emphasizing source reduction, Scott has reduced the layers of packaging in paper products for both consumers and institutional customers. It has also reduced packaging by concentrating more product inside the package; for example, by including more sheets per core of toilet tissue. Such a change may appear small, but for a major international supplier even incremental changes can have a dramatic impact.

Any company investing in paper recycling needs to understand that the business is no longer driven by demand from the paper mills. Now it will be much more of a supply-driven industry, because of the momentum of the recycling movement. The environmental movement has totally changed the complexion of the industry and opened new business opportunities. The challenge is to apply good business practices to this market and develop a more consistent supply and stable pricing environment for waste fibers. And that is precisely what paper companies are looking to accomplish. State and government incentives also play a role, but eventually it will be up to the industry to establish viable operations.

Plastics

While in-plant recycling of plastic waste is well established, recycling of post-consumer plastic waste is still embryonic. As little as 2 percent of the plastic packaging produced in the United States every year is recycled. The physical plant for recycling is still nowhere near the capacity needed, nor is application development for recycled plastics.

The strongest recycling infrastructure exists for PET; more than 200 million pounds of PET were recycled annually as of 1990. The PET bottle had reached a recycling rate of 20 percent by 1990, and it has the fastest-growing recycling rate among all beverage containers, partly because the value of used PET is second only to aluminum among packaging materials. HDPE is the second-most-recycled plastic; more than 75 million pounds of HDPE were recycled in 1990.

Since incineration is more widely used in Europe for waste disposal, the European public and activists have applied tremendous pressure on all chlorinated polymers, including PVC and polyvinylidene chloride. In the United States, major producers like B. F. Goodrich, one of the largest PVC producers in the world, and Occidental Chemical, are working to establish an infrastructure and a technology for handling various PVC products.

Polystyrene is also recycled on a relatively small but growing scale. The National Polystyrene Recycling Company has set a goal of recycling 25 percent of polystyrene used in U.S. food packaging by 1995 (see case study, Chapter 11).

Polyethylene, used in food wraps, is a high-volume material, but the difficulties in cleaning soiled food wrap are only now being resolved. Contamination from food waste also raises questions about the suitability of recycled PE for food-contact uses. One successful application of recycled PE is plastic trash bags. Webster Industries of Peabody, Massachusetts, markets the Renew and Good Sense lines of trash bags, which are made from 80 percent and 50 percent recycled plastic, respectively.

The bottle-bill states have successfully created an incentive, in the form of a deposit, for the collection of plastic bottles. This has the effect of a subsidy to support the recycling infrastructure. However, bottle bills are not consistently in force across the nation. Some have proposed a national bottle bill, but mandatory deposit laws for bottles are only a partial solution. They address only a small percentage of post-consumer plastics. And in many areas of the country, voluntary efforts have been successful in collecting the types of plastic that are not covered by deposit laws.

Many more plastic containers end up in landfills or incinerators than in recycling plants. Post-consumer plastics are plentiful; what is needed is recycling infrastructure and markets to develop and sustain consistent supply. For example, current applications for recycled PET bottles include carpeting, lawn furniture, pillow stuffing, "plastic lumber", and toys, but PET recyclers would handle much larger volumes if additional supply infrastructures were developed.

Wellman, Inc., of Johnsonville, South Carolina, the world's largest PET recycler, handles at least 50 percent of the PET bottles from the bottle-bill states, as well as PET from other sources. Wellman began collecting high-grade polyester waste from Southern textile mills in the 1950s, as part of a move to diversify from its traditional wool trading business. The company converted it into plastic pellets that could be injection-molded for such products as plastic furniture and electronic fasteners. Wellman also used the materials for nonwoven textiles such as pond and landfill liners, as well as for carpet fibers.

In the 1970s, as other major petrochemical companies began buying more of this polyester fiber waste from textile mills, Wellman saw its source of supply diminish, and it turned more and more to PET beverage bottle recycling. Today, Wellman is the largest PET recycler in the world, handling more than 70 percent of the PET bottles that the United States recycles every year. If the market creates a diversified, stable range of applications for other recycled material, it will help other companies repeat Wellman's success.

There are other technologies for dealing with PET as well. Instead of cleaning it, melting it, and making fibers or pellets, as Wellman does, other companies like Du Pont and Eastman Chemical have developed depolymerization processes, by which chemical reactions break down the PET polymer to smaller units. One of these methods, called methanolysis, uses methanol to achieve depolymerization, resulting in "virgin" quality polymer; two other processes, glycolysis and hydrolysis, use ethylene glycol and water, respectively, and achieve lower product quality at a lower cost. Hoechst Celanese has been working with Coca-Cola, and Goodyear has teamed up with Pepsi, to produce food-grade PET from post-consumer PET bottles. The Hoechst process was approved by the U.S. Food and Drug Administration for food contact applications in January 1991. In addition, Coca-Cola began testing refillable/reusable PET bottles in Europe in 1990, although the process in use there has not yet been approved in the United States by the FDA.(see Chapter 5).

Supermarkets have helped create environmental awareness, and their efforts may spur additional consumer demand for recycled products.

Indeed, package recycling depends on the informed consumer. The household that spends a half hour every week separating its trash is making an important investment in the recycling infrastructure. Supermarkets and bottle recyclers leverage this investment by collecting the household's returnables.

No single legislative, consumer, or industry initiative alone will be enough on which to build a recycling infrastructure. There must be a combination of initiatives; goodwill investments on the part of the consumer, legislative incentives, and corporate responsibility in the form of market development, infrastructure investment, and strategic alliances.

Glass

In the post-World War II period, the reusable glass bottle was commonplace. Wooden cartons full of empty glass beverage bottles were a familiar sight in the family garage or basement. The original green Coca-Cola bottle was sturdy enough to be reused an average of 20 times before it had to be crushed and reprocessed.

Interestingly, the recent resurgence of deposit laws led to a move to returnable bottles, but not reusable bottles. The reasons for the demise of the reusable bottle have to do with changes in the beverage and glass industries. Beverage distributors needed a simple, fool-proof, cost-effective system, and they found it in plastic bottles; they are unbreakable, lighter, and less expensive to manufacture and ship than glass. Having long since abandoned the reusable bottle, along with the neighborhood bottler, they are now deeply invested in a one-way system.

Glass is a costly and complex packaging material to recycle. Different colors of glass are currently sorted by hand, and this adds enormously to the cost of the process. Virgin raw materials cost less than cullet. Although it is slightly less energy-intensive to melt cullet than raw sand, the energy savings do not offset the expense of cullet. Glass plants are typically located in producer regions like Pennsylvania and Ohio, not in consumer regions, making transportation costs a serious issue. An exception is California, which has a substantial number of processing plants for both flat and bottle glass.

The reusable/refillable bottle has limits on its applications because of sanitary issues. When the packager puts a product into a glass bottle at the bottling plant, the product's integrity is intact until the consumer opens it at home. But is it possible to guarantee the same integrity when the consumer must be trusted to return an uncontaminated container? What are the implications for public health and consumer liability? Because of these uncertainties, many supermarkets that once offered refillable products have since been forced to eliminate them.

The California experience in glass recycling is instructive. The California Container Glass Recycling Campaign, established by the container glass industry, lobbied to enact Assembly Bill 2020, which established a state Division of Recycling and formalized payments made to consumers for all containers. A.B. 2020 was put forth as an alternative to other states' bottle bills, which the glass container industry saw as preferential to aluminum and plastic containers, and it leveled the playing field for competing materials.

The California law differs from other states' regulations in this way: Most bottle laws require the beverage bottler to accept returns. In the case of glass, the costs associated with transporting glass and the complexities of recycling the material often have the effect of forcing the bottler to abandon glass in favor of plastic and aluminum containers. In California, the bottler is not required to accept the returned containers. The state collects money from the packager for each container produced, and that is paid back to the consumer when the container is turned in for recycling. Consumers are paid the same amount for any container—glass, metal, or plastic—so that no one material is favored in the recycling process.

Composting

By banning yard wastes from landfills, local governments can provide a useful incentive for backyard and town composting. Composting may be regarded as another useful form of recycling, as yard wastes comprise nearly one-fifth of municipal solid waste. However, composting is not just for autumn leaves and grass clippings. In Europe, composting facilities for mixed municipal solid waste have been successfully operated for over thirty years. In Sweden, nearly one-fourth of all municipal solid waste is composted. In the United States, the infrastructure for composting is limited.

Procter & Gamble and other leading companies have formed the Solid Waste Composting Council, which is working to establish composting as an integral part of municipal solid waste systems. In Europe there are 200 composting facilities in use, and in the United States there will be nearly two dozen facilities in use by the end of 1991, and 150 more at the planning stage. Figure 4-1 depicts the composting process at a Fairfield, Connecticut, facility.

Recomp, Inc. of St. Cloud, Minnesota, is one of the most technologically advanced compost factories, turning 100 tons of trash a day into compost for Christmas tree farms, landscapers, and highway projects. An automated system separates out lead, mercury, and heavy metals before the trash is composted in as little as three days in giant metal drums.

Agripost, Inc., a Miami-based composting company, handles 800 tons of Greater Miami's garbage every day. Mixed household trash is visually inspected for lead-acid batteries, propane tanks, and other toxic or dangerous materials. The remainder—glass, plastic, metal and organic waste—undergoes a pulverizing process before it is spread out over a 6-acre indoor

Figure 4-1. An alternative to landfills: the waste composter.

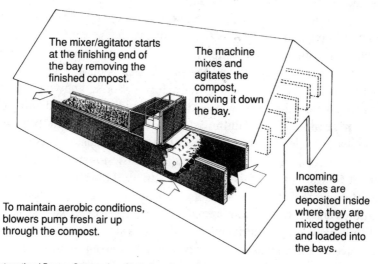

The mixer/agitator starts at the finishing end of the bay removing the finished compost.

The machine mixes and agitates the compost, moving it down the bay.

To maintain aerobic conditions, blowers pump fresh air up through the compost.

Incoming wastes are deposited inside where they are mixed together and loaded into the bays.

Source: International Process Systems, Inc., Glastonbury, Connecticut

composting facility. Microbes digest and process the carbon and nitrogen content of the organic matter, and grow and reproduce in the process. This activity is a heat-producing process, which kills harmful organisms in the compost. After a month the compost is cooled and pulverized a second time before ending up in the company's product, called "Agrisoil." Company officials say that the glass, metal, and plastic components of Agrisoil are minute, and point out that commercial potting soil often contains as much as 50 percent inorganic material such as sand, perlite, and vermiculite.

Incineration

Combustion facilities handled about 30 percent of all U.S. municipal solid waste in 1960, although most of the incinerators of that era had no pollution controls and no facilities for energy recovery. Combustion rates dropped steadily in the 1960s and 1970s, as old incinerators were closed and public awareness of air pollution created increasing demand for alternate disposal methods. In 1980, combustion reached a low of 10 percent of waste generated, but it began steadily increasing again, reaching a level of 14 percent in 1990. All major new facilities have energy recovery capability and are designed to meet air pollution standards.

The EPA projects that the incineration rate in the United States will reach about 25 percent in the year 2000. By contrast, incineration rates are already much higher abroad; for example, Switzerland incinerates 74 percent of its garbage, Japan 66 percent, and Sweden 50 percent (see Figure 4-2).

Combustion facilities exist in many forms. Some, known as mass-burn facilities, handle the entire waste stream, excluding large and bulky objects like refrigerators. The resulting ash is then deposited in landfills, where it occupies considerably less volume than noncombusted waste. In 1985, about 71 percent of the existing or planned waste-to-energy facilities in the United States were of this mass-burn variety. Other facilities, called refuse-derived fuel (RDF) operations, separate combustibles from noncombustibles and recover certain recyclable materials, especially steel.

While packaging material in general is not a significant source of hazardous emissions from waste-to-energy facilities, two emissions are of special concern: hydrogen chloride, a highly corrosive pollutant; and dioxin, an

Figure 4-2. Percentage of municipal solid waste incinerated in selected countries.

Switzerland	74%
Japan	66%
Sweden	50%
France	35%
U.S.	15%

extremely lethal toxin that is linked to cancer, birth defects, and fetal death. Some bleached paper products contain chlorine, which can generate low levels of dioxin when combusted. And PVC can produce hydrogen chloride when it thermally decomposes. In Europe, where incineration rates are higher than in the United States, PVC packaging is a major target of legislative controls, as well as of consumer boycotts.

Proper temperature controls, combined with post-combustion chambers or afterburners, can limit the formation and emission of dioxins. In addition, scrubbing systems can control hydrogen chloride and dioxin emissions as well as emissions of fly ash, smoke, carbon monoxide, and sulfur.

Since it is impractical to assume that all municipal solid waste can be recycled, waste-to-energy facilities offer a promising alternative to landfilling. Where packaging materials are concerned, paper, wood, and plastic are combustible and have a significant fuel value (see Figure 4-3). Nonfuels, such as glass, steel, and aluminum, are not combustible and, unless recovered and recycled, are usually landfilled.

In addition, banning certain recyclable materials from incineration can create an added impetus for recycling. Communities seeking to encourage composting may find that banning yard wastes from incineration will both encourage composting efforts while reducing the moisture content and increasing the incineration value of the remaining waste.

Because of the widespread lack of understanding about waste-to-energy facilities, the public generally perceives incinerators to be of the polluting,

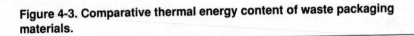

Figure 4-3. Comparative thermal energy content of waste packaging materials.

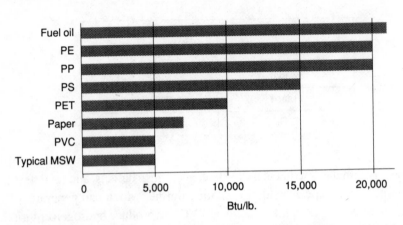

smoke-belching, 1960-era variety. The Not-In-My-Back-Yard (NIMBY) syndrome will continue to create major community resistance to new incinerator sitings. Anticipating the future increased acceptance of waste-to-energy plants, major waste management companies like Browning Ferris and Waste Management, Inc., are already acquiring and investing heavily in these facilities. Hospitals and other industries with special waste management needs are investing in their own dedicated waste reduction facilities, in what represents a new form of vertical integration. However, a major public education challenge faces both industry and legislators before waste-to-energy solutions will be accepted and adopted on a mass scale.

Landfilling

Currently over-used in the United States, landfills will continue to be an essential component of the waste management hierarchy. After as much waste as possible has been reduced, recovered, reused, or converted to energy, there will still be some waste that cannot be disposed of in any other way except in landfills. However, landfill capacity is already dwindling severely in the most populous regions of the country (see Figure 4-4). In

Figure 4-4. Remaining years of landfill capacity, by state.

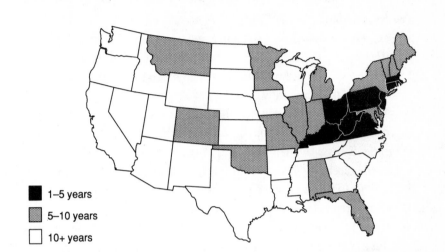

addition, the NIMBY syndrome will make future landfill sitings problematic. As a result, landfill space is an increasingly precious commodity.

Today's sanitary landfill is a far cry from the open dumps of the pre-1980 era. Those eyesores were breeding grounds for vermin, and the practice of burning trash contributed to water and air pollution. Federal legislation led to the phasing out of all open dumps by the early 1980s. Burning was banned, and deposited waste required a daily cover of six inches of soil.

However, new problems arose, including the leaching of contaminants into groundwater. Because it is difficult to detect and to clean up, groundwater (aquifer) contamination is a more intractable problem than surface water pollution, for which natural cleansing mechanisms exist. As a result, municipal landfills are now required to have bottom liners to prevent leachates, as well as surface caps of soil and other materials to prevent water from entering and percolating through the landfill, carrying toxins to water sources. Leachate collection systems are an additional measure to collect and treat contaminants before they reach groundwater.

Though degradation is extremely slow, landfills also emit methane and other gases. The methane emission from landfills has a heating value of 476 Btu per standard cubic foot. In 1988, as many as thirty-three facilities collected methane for fuel from sanitary landfills, and twelve of these upgraded the methane to a higher Btu value by removing some carbon dioxide.

The EPA's stated landfill objectives are to reduce the volume of waste in landfills, extend the landfills' useful life, make the waste more benign, and increase the landfills' safety for humans and wildlife. Industry has developed some remarkable innovations to help achieve these objectives.

3M Co. and a small company called Rusmar, Inc., based in West Chester, Pennsylvania, have both developed landfill sealing foams that can effectively seal landfills without taking up room. Rusmar's foam is made of harmless foaming agents used in shampoos and cosmetics; 3M's "Sanifoam" uses formaldehyde, urea, and other common chemicals that are also considered harmless. The foam provides fire, odor, and vermin control, and, most importantly, extends the landfill's useful life.

Other landfills are experimenting with plastic tarpaulins that can be rolled over the landfill at the end of each day and rolled back in the morning. Others have explored so-called geotextiles; these are thin, biodegradable fabrics that can be rolled over the trash daily and covered with new trash the next day.

Experiments are also underway to investigate the benefits of deliberately accelerating the process of decomposition in landfills by a controlled increase in moisture content. Early results indicate that when this moisture is recirculated, the potential for leakage is reduced and the landfill's decay period is reduced from decades to a few years. If this work is successful, it could open the way to a new generation of landfill technology.

While landfills remain necessary and pose no immediate problem to less populous states where space is available, environmental principles demand that society make better use of the energy and raw material value of its waste than to bury it, at increasingly higher costs, in the ground.

Degradable Packaging

When trash is compacted and covered, very little decomposition occurs in landfills. The rate of decomposition depends on a number of factors, including moisture, temperature, and permeability of the soil cover. In all cases, the decomposition process in a sealed landfill is extremely slow and can take many decades to achieve.

Most consumers are surprised to learn that their degradable waste does not "return to nature" in a landfill. The "Garbage Project," led by archaeologist William Rathje of the University of Arizona, has studied sealed landfills in Tempe, Arizona, and several sites in New York. This research has shown that virtually nothing decomposes in a landfill over as long a period as three decades. Legible newspapers and still-recognizable hot dogs and pastries were among the more unsettling finds of Professor Rathje's landfill research teams.

The two most recognized and easily achievable forms of degradation are consumption by microorganisms, or biodegradation, and deterioration by ultraviolet light, or photodegradation. For some materials, hydrolytic degradation, resulting from the action of water, is also important.

Paper is a biodegradable material, although its rate of degradation in a sanitary landfill is slower than generally realized, as Rathje's work has demonstrated. In addition, wax or plastic coatings will hinder or slow paper's degradation rate still further. Steel and aluminum are inert to microbial action but are subject to oxidation. Here again, coatings can slow oxidation, preserving cans and foils through years of exposure. Glass is highly inert; although it is subject to hydrolytic degradation, the rate is so slow as to be meaningless.

Plastics are basically stable, but can be made photodegradable, biodegradable, or hydrolytically degradable depending on their composition. For example, cornstarch has been combined with polyethylene to create biodegradable plastic. It is actually the starch that breaks down, and when it does, the plastic loses its structural integrity.

Because degradability is so difficult to achieve in landfills, its benefits are minimal except in a few selected circumstances. Degradability can play an important role in agricultural mulch. Ring-type carriers for beverage six-packs have been a successful application of degradable plastics; because they eventually lose their structure when exposed to the elements, they reduce the threat to wildlife.

Market Consolidation
and Alliances

The packaging and waste management industries have seen a number of joint ventures and strategic alliances in recent years, especially between waste companies and plastic manufacturers: for example, Du Pont, Mobil, Amoco and Alcoa on one side, and Waste Management Inc. and Browning Ferris Industries (BFI) on the other side. Recycling giant Wellman established an alliance with BFI to ensure a continuing supply of PET bottles. Such arrangements are logical and have the effect of simplifying the infrastructure by moving large volumes of material from one party to another.

The size of the players in the infrastructure raises interesting possibilities. There are two opportunities for industry: that of the large conglomerate, and that of the small entrepreneur. On the one hand, the infrastructure may work better when larger companies acquire and control larger sections of the distribution system. On the other hand, recycling will be an area of attractive opportunities for entrepreneurial companies for some time. At a time when the system is marked by consolidation and both horizontal and vertical integration, the activity of recycling is growing at such a fast pace that there will be expanding opportunities for startups that can identify profitable niches.

An example of an entrepreneurial company that has identified a waste management niche is Re-Source America, based in Southampton, Pennsylvania, which specializes in source reduction of packaging materials as a service to product manufacturers. The company recovers packaging materials from end users and inspects and/or refurbishes reusable packaging components for resupply back to the product producer. The company recycles what is not reusable to keep the packaging out of the municipal solid waste stream, thus providing a closed-loop reuse program and recycling program.

One of the incentives for forming infrastructure alliances would be to gain better control of the supply of material. Recyclers are at the mercy of the cyclical nature of the materials market. The price of plastics, for example, rises and falls like a rollercoaster and the profitability of the recyclers parallels the wild swings in scrap material prices.

This price volatility may be the most serious problem recycling faces. For a healthy infrastructure, there may have to be subsidies or price supports to allow businesses to ride out the rollercoaster business cycles—especially the bottoms—and to make sure good habits are not broken. The recycling business should no longer be at the mercy of material economics. There is now the question of waste disposal economics and landfill space, and the need to get usable material back into the economy.

Waste Collection

Municipalities currently handle about 30 percent of all waste and own most of the nation's worst landfills. Most private waste collection companies are small, local trash haulers, which handle 42 percent of garbage collection. As regulations for waste disposal become more stringent, the small trash collectors will be hard pressed. The trend is toward large companies like Browning Ferris Industries, Waste Management Inc., Laidlaw Waste Systems, Attwoods PLC, Western Waste Industries, and Chambers Development to continue to acquire more of these local trucking and hauling operations and bring them under the umbrella of major private corporations. In the fiscal year ending September 30, 1989, BFI alone had acquired 131 small companies.

Municipal solid waste generally is the largest home of privatization in the environmental business, and the trend is toward continued privatization. As municipalities study the need for a new infrastructure, the greatest hope is in the increased involvement of the private sector.

This does not mean that some states and some municipalities will not be capable of well-run operations, with advanced methods for sorting, selling, and marketing municipal solid waste. From the municipality's point of view, the ability to "go it alone" is a bargaining chip in dealing with private groups. Municipalities can threaten to take back the business if private companies do not perform as expected.

Some municipalities, especially the large cities, have very efficient and well-managed public works departments. For the most part, however, municipalities need assistance: financial, organizational, and instructional. Private involvement will be significant, although waste management will not be 100 percent private.

Incineration raises infrastructure issues that are worth exploring. The waste-to-energy business has a very important role to play in the infrastructure. Even if the objective of recycling 50 percent of all waste material were reached, there is still the dilemma of what to do with the rest.

If we recycle more plastics and paper, we have less "fuel" to burn for waste-to-energy facilities. At one time this was an intensely debated issue. Now the waste companies know they have no choice: They are being told to recycle, or else lose out on gaining municipal contracts. In the end, the waste companies will need to reckon with broad societal pressures for recycling, reduced packaging volume, incineration, and composting.

National Players, Regional Decisions

State governments can create an important driving force for recycling when they dictate a preference for recycled materials in their procurement policies; that creates a new market opportunity for recycling. Where legislation falls short is in the inconsistencies among different states and different regions. The states that do not have bottle bills tend to be those whose landfill capacity is not an immediate issue. Without consistent legislative mandates, recycling efforts are concentrated in some regions, diluted in others.

A consensus needs to be developed among many players: state legislatures, consumers willing to deal with the inconvenience of separating their trash, supermarkets willing to tolerate the nuisance of collection, and businesses with an eye to the opportunities in the emerging infrastructure.

Notes

1. Susan Selke, *Packaging and the Environment* (Lancaster, Pa: Technomic Publishing), 1990, p. 90.

The International Perspective

5

It is no coincidence the world's most highly industrialized regions—
Europe, Japan, and the United States—are also home to the most significant
environmental movements. Nowhere is this more true than in western
Europe. Companies based outside of Europe are often amazed by the European public's intense concern over environmental issues.

In contrast with the United States, which accommodates a variety of
environmental movements and many grass-roots factions, the European
environmentalists have progressed far beyond the grass-roots stage to
achieve broad political power and unity through "Green" parties. Legislators and businesses throughout the world look to Europe, especially to the
Netherlands and Germany, as bellwethers for future environmental trends.
(It may also be noted that the Netherlands and West Germany, along with
Canada, devote a larger proportion of their gross national product to environmental expenditures than does the United States. However, some other
European countries spend less. See Figure 5-1.)

Packaging, too, is directly influenced by national characteristics. The
United States developed transportation and distribution systems to move
everything from fresh produce to delicate instruments across a 3,000-mile
span. To protect these goods over long distances, the United States devel-

Figure 5-1. Average annual environmental expenditures, as percentage of GNP, selected countries, 1980 – 1988.

Netherlands	1.34 %
Canada	1.11
West Germany	1.07
United States	0.87
United Kingdom	0.74
Japan	0.69
Sweden	0.64
France	0.62
Italy	0.55

Source: Organization for Economic Cooperation and Development

oped advanced packaging concepts that are second to none. In addition, American culture gave rise to fast-food restaurants, supermarket shopping, and frozen foods, each with its own implications for packaging. By contrast, the traditional ritual of daily shopping for fresh food persists in much of the world, including Europe and Japan, and has meant very different concepts in food packaging. And in Japan, packaging has an almost ceremonial aspect that is unknown in the West.

Clearly, culture and geography are key determinants of environmental consciousness, and they define the role of packaging in society. This chapter will focus on Europe and Japan: the history of their environmental movements, and the packaging industry's response.

The Greening of Europe

The Green Movement in Europe has a complex history. Like many modern grass-roots movements, it started in the 1960s and was a reaction to 20 years of massive postwar growth in industrial output, particularly in West Germany, where the environmental movement became the most publicized and the subject of historical and political analysis.

The West German Green Movement was an outgrowth of West Germany's Citizens Initiative Movements, which started in the 1950s. These were local/regional pressure groups formed to stop, or at least delay, the excesses of industrial infrastructure development that damaged the local ecology. The local initiatives grew during the 1960s and led to the first phase of the Green Movement. During this phase, the movement was not questioning the sociopolitical systems, but was protesting on a project-by-project basis— opposing a nuclear power station, a new airport runway, and so on.

By the early 1970s, the movement was closely identified with ecology in all its aspects, including housing and city and traffic planning. The energy crisis of the 1970s, which accelerated the political commitment to nuclear power, propelled the ecology movement into its second phase, in which it truly gathered momentum and became increasingly system-critical and system-opposing both in theory and practice. This led to the Green Party's political beginnings, as its proponents began to understand the workings and interconnection between government and large-scale industries.

At this stage the Anti-Nuclear-Energy Movement and the Ecology Movement were running side-by-side in West Germany, both advocating a halt to economic growth and, hence, both totally opposed to the government's economic policies. This activity paralleled the growth of Green movements in other Western European countries, but these had lower profiles, mainly because of slower economic growth patterns prevailing there.

Interestingly, Eastern Europe had been going through a similar industrialization process, but protest was submerged as a result of complete state control. The dire results of this lack of "people pressure" has only recently been fully realized, and it will emerge as one of the top challenges of a united Germany.

The Ecology Movement's criticism of industry and technology was aimed primarily at their polluting and destructive effects on the environment. Later, the question of depleting limited resources in a finite world was an added focus of the Greens. This led to the third phase of the movement, where industrialism itself was criticized and a "back-to-nature" approach gained prominence, under the rallying cry, "There is no such thing as harmless high technology." Although the movement could delay many projects, it could prevent very few, and from this stemmed the desire for more political clout. In 1980 the Green Party was formed in West Germany. Die Grünen won seats in most local councils and state parliaments and in 1983 entered the Bundestag, after having won 5.6 percent of the vote.

The final phase of Die Grünen's development has seen its political maturing and its linkage to the New Peace Movement, aimed at scrapping nuclear missiles in Germany. This has led to a parting of the ways between some grass-roots factions and the political party. Die Grünen has had to compromise to stay alive politically and is therefore seen as the "establishment" by more extreme wings of the movement. These internal squabbles have cost the party some credibility. In the December 1990 elections, the German Greens mustered only 3.9 percent of the vote, less than the 5 percent needed to secure a parliamentary seat, and well below the 8.3 percent they had won in 1987.

While Die Grünen enjoyed the greatest publicity and widest electoral success, it was the Belgian Greens that first gained national representation, in 1981. In fact, Ecolo (the French-speaking Belgian Greens) and Agalev (the Flemish-Belgian Greens) had a combined national vote above 10 percent by 1990. By comparison, the Green parties in the United Kingdom and France have lagged behind, although the French Green party has grown strong especially at the local level; several French cities, including Strasbourg, have elected Green mayors. The British Green Party began as the Peoples Party in 1973, became the Ecology Party two years later, and the Green Party as late as 1985. It has yet to win real representation at a national level but has been gathering strength.

The growth of these political parties paralleled a number of environmental catastrophes in Europe. In 1976, a major accident at a chemical plant in Seveso, Italy, released a cloud of deadly dioxin. The 1986 meltdown of a nuclear plant in Chernobyl, in the Soviet Union, not only cost a heavy toll within the Soviet Union but also deeply affected Sweden, Switzerland, Germany, and Austria, where crops and livestock had to be destroyed. A 1986 fire at a Sandoz Chemical warehouse in Basel, Switzerland, released large quantities of mercury into the Rhine. These and many other local disasters have sensitized Europeans to the more global problems of ozone depletion, acid rain, tropical deforestation, and the greenhouse effect.

It is generally agreed that the turning point in the politics of the environment was Margaret Thatcher's speech to the Royal Society in September 1988. It was significant that, prior to that moment, no single national leader of global stature had devoted the bulk of an address to the problems of the global environment. Since that speech, nearly every national leader has made speeches and taken initiatives on the environment.

The growing awareness of environmental issues is shown in the political support more recently shown for the Greens in much of Europe. In the 1986 general election in the United Kingdom, the party had polled only 1.4 percent of the vote. But in the elections to the European Parliament in 1989, Greens polled 15 percent of Britain's vote, thirty times more than in the 1984 elections to that body. In the whole European Parliament, as a result of the 1989 elections, Greens and their allies won 39 of the 518 seats, nearly doubling their representation there. This was very much a protest against the policies of existing political parties in power. All parties, whether to the left or right, now have the environment as a major policy plank, many jolted into action by the size of the Green vote. Environmental regulatory programs are highly or at least moderately developed in most major countries of Western Europe (see Figure 5-2).

Figure 5-2. Development levels of European environmental regulatory programs.

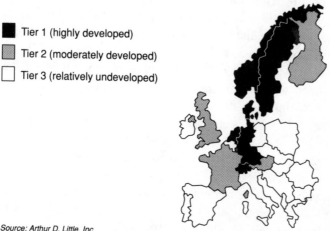

■ Tier 1 (highly developed)

▨ Tier 2 (moderately developed)

☐ Tier 3 (relatively undeveloped)

Source: Arthur D. Little, Inc.

The Green parties in each country have had the desired effect, but politically they have worked their way out of a job. Because of the narrowness of their fundamental approach, none will achieve long-term political power. However, the principles behind the Green Movement are here to stay, woven into the fabric of other political parties. As with all such societal movements, first it is ignored, then it is attacked, and then the sensible part of its message is absorbed by others under the pretense that they had been thinking about it all the time. The Green Movement is now in this last stage.

Legislative Impact of the European Community

In recent years the pressure for improved regulations has increased tremendously. The European Community (EC) environmental policy dates back to 1972 and has since developed through "Environmental Action Programs." As a result of these programs, the EC has taken policy initiatives across a wide range of environmental issues including water, air, and noise pollution; waste management; land use planning; and wildlife protection. The EC's declared policy for 1992 is to make the Community's environmental standards approximate the highest of the levels existing at the time among member states, although member states remain free to impose stricter regulations.

Since the inception of EC policy, well over 100 proposals have become law in the form of directives. Directives are legally binding on the EC member states as to the result achieved, but the form and method of implementation remains within the discretion of each government. An example is the EC's 1982 Seveso directive, which was inspired by the 1976 dioxin disaster in Seveso, Italy; interestingly, it later became the model of the Superfund Amendment and Authorization Act (SARA) in the United States.

Until recently, the primary justification for the EC's developing a common environmental policy has been to prevent distortions in competition caused by the unequal production costs associated with different national environmental measures. In addition, these national environmental measures have been perceived by many as thinly veiled nontariff barriers to trade, interfering with the free movement of goods within the EC. Although the role of the EC in protecting the environment for its own sake was explicitly recognized for the first time in 1987 amendments to the EC

Treaty, economic considerations are still a powerful force in environmental policy.

Waste management has been an integral part of environmental policy since the mid-1970s. Key principles in this area have been:

- Proper management of natural resources
- The "polluter pays" principle
- The idea that prevention is more efficient and less costly than cure

Consistent with these principles, EC policy has always emphasized waste minimization, reuse and recycling as preferable alternatives to landfilling or incineration.

By 1974, after various member states had taken legislative action to control waste, the EC Commission (the Community's executive body) saw the need for a directive to harmonize national legislation and create a coherent pan-European system for waste management. The result was the 1975 Directive on Waste. While this provision established a framework for EC measures on waste management, it did not give the Commission any enforcement powers to compel member states to reduce the waste stream. The volume of waste and the difficulties in safely disposing of it have continued to grow, with the result that the Commission has looked increasingly to obligatory recycling.

Much more effective was a "daughter" directive, put forth by the Commission in the early 1980s, to reduce the number of containers for consumable liquids in the waste stream. Somewhat surprisingly, this directive was adopted, after substantial amendment, by the Council of Ministers in 1985. (The Council, designed to protect the interests of individual EC nations, is composed of representatives from each government and usually reaches decisions unanimously.)

In accordance with this step-by-step approach, the Commission has also considered action on waste paper, compostable waste, and waste plastics, especially from plastic packaging. The Commission has established ambitious goals to reduce the amount of waste plastics going to incineration or landfill by as much as 80 percent. These are examples of the substance-based, rather than product-based, approach now favored by the Commission. They are likely to be followed by other substance-based proposals on, for example, the use of metal beverage cans.

The Commission has also put forth an EC "eco-labeling" proposal. Eco-labeling was introduced in West Germany in 1977; the country's Environmental Agency uses the United Nations Environment Program's "Blue Angel" seal of approval for environmentally friendly products, from phosphate-free detergents to reusable bottles. In February 1991, the EC announced a more comprehensive eco-labeling scheme that would only be granted to products satisfying stringent criteria, based on life-cycle analyses of their manufacturing processes, use, and disposal. The analyses would examine the amount of natural resources and energy consumed, the waste and noise pollution produced, and the tendency to cause air, water, or soil contamination. Use of the EC's eco-label, which is voluntary and may run for a three-year period, is likely to give a commercial boost to any product. Food, drinks, and pharmaceuticals are excluded.

In November 1989 the EC voted to create a European Environmental Agency (EEA) that would begin work in 1991. The European Parliament had hoped to model the EEA after the U.S. Environmental Protection Agency, with broad authority to enforce regulations. However, because some member states wanted to retain enforcement powers, the EEA is initially a central information clearinghouse and a coordinator for national centers of environmental monitoring and evaluation. Once it is well-established, the EEA may acquire greater authority to investigate compliance with EC environmental laws.

Beverage Containers Directive

The 1985 EC Directive "on containers of liquids for human consumption," now loosely described as the beverage containers directive, would, in its original form, have tied member states to specific targets for the refilling and recycling of containers and for the reduction of containers in the household waste stream. However, successful lobbying by industrial interests, as well as the fact that different drinks are popular in different cultures—for example, tea in England, beer in Germany, wine in France—softened the measures that eventually were agreed upon.

The directive requires member states to "draw up programs for reducing the tonnage and/or volume of containers in household waste to be finally disposed of." Member states are obliged, either by legislative or administrative means or by voluntary agreements, to take measures designed to:

- Develop consumer education on the benefits of refilling and recycling;
- Facilitate refilling and recycling;
- Promote selective collection of nonrefillable containers and recovery of materials;
- Encourage technical development and marketing of more "environmentally friendly" containers;
- Maintain and, where possible, increase the proportion of refilled and recycled containers; and
- Ensure that new refillable containers are clearly marked as such.

A primary impetus for this directive was to prevent the fragmentation of the European market that would result if individual member states introduced measures that would effectively restrict imports. However, while the beverage containers directive did not require member states to promote refilling or recycling, it did not appear to rule out such measures. As a result, Denmark decided to continue with its ban on nonreturnable bottles for beer and soft drinks, subject to some minor modifications. Several other member states were quick to propose mandatory schemes of one kind or another. West Germany proposed to institute a compulsory deposit of 0.5 marks on all plastic bottles (at a time when the deposit on refillable glass bottles was 0.3 marks) and require the distributor to take the bottles back and guarantee they would be recycled.

Italy introduced proposals for a consortium to recycle containers, to be financed by a levy on materials, and proposed tax penalties for failure to meet recycling targets. The Italian law would also require all packaging to be biodegradable or recyclable, although the law was contested before the European Court of Justice as an unlawful restriction on trade.

Ireland proposed a ban on cans and PET containers for beer, cider, and wine, as well as a ban on ring-pull cans for soft drinks. These ideas were, however, rejected by the Commission as disguised barriers to trade.

Such proposals were viewed with alarm by the beverage manufacturing and packaging industries and by several member states. Industrial interests, particularly those concerned with plastics and cans, argued that many of the measures were introduced to erect trade barriers rather than for environmental reasons and urged legal action against the countries concerned. At the same time, however, manufacturers and their associations, in anticipation of further EC legislation, began to develop their own recycling initiatives.

From the perspective of environmental protection, the Commission welcomed measures taken by member states such as Denmark and Germany. Meanwhile, other member states, including the United Kingdom, were doing very little to promote reuse and recycling, demonstrating the directive's ineffectiveness to improve performance across the EC. The difficulty for the Commission overall, however, was that measures such as those being pursued by the Danes and the Germans appeared likely to cause fragmentation to the common market and disruption to the free movement of goods.

These concerns led the Commission to make a landmark decision with far-reaching implications for manufacturers and users of packaging. The Commission resolved to take Denmark to the European Court of Justice, charging that the ban on nonrefillables breached Article 30 of the EC's founding treaty. Article 30 prohibits restrictions, or measures having the effect of restrictions, on imports from other member states. The Commission intended this to be a test case which, if successful, would also halt comparable laws in other member states.

The Danish Bottles Case

The 1981 Danish law required all beers and soft drinks to be sold in refillable bottles bearing a mandatory deposit. Metal cans were banned, and the requirement that containers be reusable also effectively prohibited plastic bottles. The law also required containers to be approved before use in order to standardize bottles and make the universal return scheme more effective.

This scheme proved highly successful in environmental terms, with a 98 percent return rate being achieved. However, the beverage industry outside Denmark complained to the European Commission about the barriers to their imports that the law created.

Under EC law, national laws that create barriers to trade are permissible under certain circumstances—for example, if they are deemed necessary to satisfy "mandatory" requirements of EC law. The EC Court concluded, first of all, that protection of the environment has been one of the declared objectives of the EC since the Single European Act of 1986, and as such is a mandatory requirement of EC law. Secondly, the Court ruled that obligation to use refillable deposit bottles was "an essential element of a system aiming to secure reuse of containers and therefore appears to be necessary to attain the objectives of the disputed law."

Reactions to the Court's decision were predictable. Environmentalists and member states hoping to introduce similar measures were delighted. However, the drinks and packaging industries feared more restrictions and further fragmentation of the market. The EC Commission, while pleased on its environmental side, was more generally disturbed about the setback to free movement of goods and harmonization of laws.

The Court's judgment was delivered in September 1988. Later, the Commission signaled its intention to come forward with new legislative proposals, but this did not stop Germany and Italy from introducing restrictive measures of the kind they had earlier proposed, soon to be followed by Belgium, Holland, and Ireland. Germany, for example, in addition to introducing its mandatory deposit on plastic bottles, stipulated that by mid-1991, 90 percent of beer and mineral water containers, 80 percent of carbonated soft-drink containers, 50 percent of wine bottles, and 35 percent of noncarbonated soft-drink containers must be returnable, failing which mandatory deposits on one-way containers seems inevitable. Meanwhile, industry has been lobbying against new and more restrictive EC measures and is busy preparing its own voluntary recycling schemes in the hope of persuading the Commission that they offer a satisfactory alternative.

The Court's judgment in the Danish bottles case was significant in demonstrating the Court's willingness to protect the environment, even where this may interfere with free trade. This decision may have serious implications for the EC's future economic and environmental policies, especially for the beverage and packaging industries.

Green Consumerism

In addition to legislation, the rise of the "green consumer"—the legacy of the ecology movement—is putting pressure on industry to change its ways. In this debate, two strands of "greening" are coming to the fore; there are "deep greens," who feel that society should be consuming less, and "shallow greens," the majority of the public, who feel society should be consuming better by buying products that use less energy, less resource, and less packaging.

The underlying pressure to become Green consumers is pervasive. Witness the best-selling paperback, *The Green Consumer Guide*, in the United Kingdom, the "green label" approach of many large European supermarkets, and national eco-labeling programs such as Germany's Blue Angel system.

Retailers have been first to feel the impact of green buying power. Tesco, a British supermarket chain, made Greening a central part of the selling pitch, and used environmental friendliness to shift its image upmarket. Tengelmann in West Germany stopped selling CFC-based aerosols in 1988, and in 1990 it told suppliers that all products and packings containing cellulose must be chlorine-free. Migros, Switzerland's biggest retailer, has a computer program to check the "eco-balance" of packaging: an application of life-cycle analysis to determine which kinds of packaging take the most resources to make and to dispose of. IRMA, one of the largest supermarket chains in Denmark, has committed itself to providing "ecological" products, from range-fed chickens to organic produce, and has banned the use of PVC packaging from May 1990.

The most quoted example of "green" retailer success is the United Kingdom-based Body Shop, started from nothing by Anita Roddick. The Body Shop sells personal products that are "cruelty-free" (not tested on animals), minimally packaged, and based on natural ingredients, and it has been continuously successful even during the current retail recession. Its message and outlets are now seen all over the world. Most consumers see this as "good," but is this "green product" growth adding to, or replacing, existing consumption? A definitive answer is not yet available.

Green consumerism has spawned a new industry of its own: "green publishing" (using recycled paper, of course). Titles such as *The Green Capitalists* and *Green Pages: The Business of Saving the World* roll off the presses in increasing numbers.

The Green Capitalists shows how industry is incorporating environmental considerations into profitable growth. Examples given are cleaner, quieter, more fuel-efficient Rolls Royce aero-engines; Mercedes-Benz's use of catalytic converters before legislation mandated them; and ICI's development of a car paint that cuts emissions by 85 percent during drying. The extension of this latter approach is Toyota's use of water-based paints in its new British car factory in deference to environmental concerns. It seems that the large companies are now taking the offensive. By becoming world leaders in the development of less-polluting car paint, not only are car manufacturers reaping large profits, but the fundamental question about whether we should be manufacturing more and more cars in the first place is pushed down a few notches on society's political agenda!

Waste Management Options

In the United States, a typical community with a population of half a million and producing 2,000 tons a day of waste currently, will require a 240-acre landfill by the year 2010. Most European countries would be hard-pressed to devote such large parcels of land to landfilling, and may not need to. The current waste generated per person in the United Kingdom is one-third that of America. But as convenience packaging becomes more and more popular in Europe, the waste problem will become increasingly severe.

Landfill sites across Europe are increasing "gate prices," or tipping fees, to cope with the increase in waste already and the lack of availability of suitable landfill sites. As we have seen, the EC has established ambitious goals to reduce the amount of waste plastics going to incineration or landfill. Minimum recovery and recycling percentages are to be set at a "fairly high level," as high as 60 percent in the case of packaging and domestic wastes. In addition, responsibility for collecting wastes is to be defined and may involve businesses at any point in the chain from production to material recovery. Other measures to promote recycling include a value-added tax (VAT) abatement for recycled plastics and penalties for noncompliance, to include forcing businesses to store unrecycled waste until facilities are provided.

In the light of EC law following the Danish bottles case, there are severe limits to the concessions that can be made to industry without encouraging new national restrictions, leading to increased fragmentation of the common market. This is likely to be a sobering thought for the European packaging industry. The only viable solution may be for European companies to do as their U.S. counterparts have done and embrace post-consumer recycling on a large scale as a profit-making venture. There now are signs that industry itself is thinking along these lines.

New guidelines are being written as amendments to EC directives, such as the directive on liquid beverage containers. Within three years of the adoption of this directive, 70 percent of liquid food packaging should be recovered for refilling or recycling; 80 percent after six years. If deposit systems are imposed to meet the targets, the system must apply to all forms of packaging used in that sector. Incineration with energy recovery is included in the definition of recycling, with no limits on this form of recycling as compared with material recycling. Thus, a recycling requirement could, according to this test, be entirely fulfilled by "energy recycling" in locations where incinerators are appropriately equipped (see Figure 5-3).

Legislation throughout Europe will regulate plastics, but will not always differentiate between competing materials. Moreover, all parties involved in manufacturing, selling, and disposing of plastics will be obliged to participate in some way in providing real solutions to the municipal solid waste problems.

EC directives are also being considered for all containers, for all types of packaging and labeling, and for landfill sites. Under such legislation it would eventually be required that 70 percent of all packaging materials be refillable or recyclable or be able to be incinerated in a process that provides

Figure 5-3. Materials and energy recovered from MSW in some European countries.

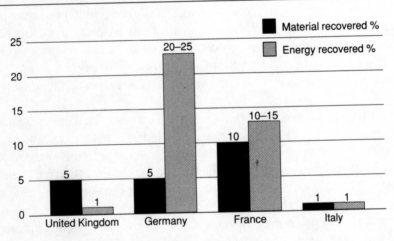

Source: Environmental Resources Ltd. Resources Recovery

for energy recovery. When this legislation will eventually be implemented is indefinite, due to the ever-moving goal posts associated with pollution legislation.

Energy from waste plants will have to meet stringent emissions standards to comply with future legislation. Although the standards can be met, the costs will be high. This means that small-scale plants will become less economically viable and larger, more centralized schemes will increasingly be developed.

It is predicted that landfill costs will quadruple in some areas over the next few years because of a lack of suitable sites and a tightening of regulations controlling them. Packaging manufacturers and local governments alike will be under increasing pressure to explore new waste management options.

Packaging Industry's Response

What has been the response of European industry to these multiple pressures from activist groups, legislation, and consumers? The packaging industry has been faced with several problems at once: an image problem with consumers, a record of inaction, and a lack of knowledge on the part of the public. The industry's image and credibility have suffered partly because companies are seen as having overpackaged products in order to differentiate them in the marketplace. The industry has been slow to act on its own highly developed awareness of the total effect of its products from "cradle to grave" on the environment. Finally, industry needs to educate a public that does not know the intricacies of preservation techniques that reduce spoilage and hence the product waste within the package.

In an effort to offset the public's poor conception of the industry, national associations of manufacturers have been formed to disseminate information about packaging and the manufacture and retailing of packaged goods. A good example is INCPEN, the Industry Council for Packaging and the Environment. Founded in the United Kingdom in 1974, it is now being used as a model in other European countries. The aims of INCPEN are to further the protection of the environment insofar as it is affected by packaging, to promote the environmental benefits of packaging, to protect member interests against unjustifiable attack, and to inform members of political and social developments affecting packaging and the environment.

Packaging companies are putting environmental strategy at the highest level of priority. The British chemical giant ICI pledged to double its environmental spending between 1990 and 1995, to establish recycling programs for its products and packaging both in-house and with suppliers, to step up resource and energy conservation, and to cut wastes by 50 percent. Three German plastics companies—BASF, Hoechst, and Bayer—established a joint recycling venture in 1990, called EWK, to promote recycling research. Hoechst built a recycling plant near Cologne with an annual capacity to process 5,000 tons of polyolefins.

The Swedish operations of Tetra Pak, makers of food and drink cartons, rely on a paper mill that bleaches wood pulp without chlorine and derives raw material from meticulously managed forests. Forest growth in Sweden exceeds tree fellings by 3–5 percent annually; some 80 percent of Sweden is covered by forest, nearly all of which was planted in this century.

Coca-Cola launched a successful program of returnable PET bottles in Switzerland, Germany, the Netherlands, and Norway, in addition to its initiative to use recycled content in PET bottles worldwide. The returnable and refillable PET bottles use technology developed by a Swiss company, Soudronic SA, including ultraviolet light and a "sniffer" to detect contamination.

And in the Netherlands, which is becoming something of a recycling center in Europe, the Reko recycling subsidiary of DSM has grown an average of 25 percent a year and has plans to double its capacity in film, sheet, and board from recycled material to 8,000 tons per year. Material generated from this investment will be used in construction, agricultural, and automotive applications in the form of finished products. This will bring Reko's overall recycling capability to around 30,000 tons a year, making it a world leader. Its success has been exported to North America via a licensing deal with Johnson Controls for PET bottle recycling technology.

The Future Partnership

With the high level of media coverage on macro-environmental issues like ozone depletion and global warming, the future of our planet is rapidly becoming a worldwide concern. Although general governmental agreements can be responsible for sweeping short-term changes—such as chlorofluorocarbon (CFC) replacement in aerosols, refrigerators, and so forth—

the consumer has a major influence when it comes to environmentally friendly packaging. It is therefore particularly important that a team or partnership approach is taken when considering how packaging can really become environmentally responsible from the standpoint of waste management, resource use, recyclability, or reusability.

The partnership requires enthusiastic collaboration between government (local, national, and international), industry (from raw material supply to finished consumer goods supplier), and consumer. This partnership is building in Europe around a framework of tighter legislation and standardization. However, if it is going to work, it must be grounded in community partnerships and support.

The great majority of European countries have landfill shortages similar to that of the population centers of the United States, with the large urban centers having the greatest worries. The recent British government white paper on the environment entitled "This Common Heritage" commits the United Kingdom to meeting a target of recycling 25 percent of its domestic waste by the year 2000, as against a current figure of 10 percent. As part of this policy, the British government plans to:

- Work with local government to assess the effectiveness of experimental recycling programs.
- Encourage more recycling "banks" for cans, bottles, and plastics.
- Create a labeling system for recycled and recyclable products.
- Look for other ways to encourage companies to recycle more building materials and mining waste.

All this is built into a two-pronged strategy based on regulation and fiscal incentives—in other words, the "polluter pays" principle.

Germany, which always looms large in the environmental debate, introduced mandatory deposits on PET bottles and a scheme that would require German packaging manufacturers to take back all packaging materials passing through retail outlets for recycling. German consumers were also encouraged to strip off all unwanted packaging at the supermarket, with the responsibility for disposal placed in the retailer's hands. The federal cabinet imposed an 80 percent recycling target for plastic, glass, and paper packaging by 1995; however, plastics producers BASF and Hoechst protested a concurrent ban on incinerating plastics for energy recovery, an option which they believe is "indispensable" and "ecologically desirable."

In the Benelux countries, governments are becoming increasingly active in waste reduction; this will have the effect of shifting increasingly to packaging materials seen as environmentally friendly, such as glass and tinplate, which have a high recovery rate. The Netherlands is leading the way in European recycling; because it is one of the most densely populated countries in the world, the resulting pressure on landfills has made recycling a virtual necessity. The glass recycling rate is around 50 percent, and wastepaper collection in the schools is a part of the curriculum. PET bottles are extremely popular, and pressure groups have made PET recycling a top priority.

While the level of recycling in this region is already high by European standards, the Benelux countries have set ambitious targets to increase the future reuse and recycling of waste materials, including packaging. The Dutch government's National Environment Plan, introduced in 1989, set out stringent targets for packaging waste. They are:

- To reduce the total waste stream by 10 percent between 1986 and the year 2000.
- To avoid materials such as heavy metals and CFCs.
- To increase the reuse and recycling of packaging waste from 25 percent to 60 percent.
- To increase the incineration of packaging waste from 25 percent to 40 percent.
- To reduce the share of waste dumped in landfills from 50 percent to zero.

These goals are typical of northern Europe's approach to waste management. Southern Europe still maintains a routine of daily shopping in all but the largest cities, and therefore the highly packaged, highly preserved food syndrome is less extensive. However, the proliferation of snack food products and fast food outlets are the starting point for a future waste management problem in the southern European countries.

In response to the EC's emphasis on recycling, some industry leaders have responded that the collection, transport, and cleaning processes associated with recycling have environmental impacts, and that the overall effect may be negative. The most promising recycling efforts are regional or local recycling partnerships, which offer the opportunity for some immediate return from the recycling process.

Local government's role is usually to facilitate the collection and disposal of household waste. Local schemes are being developed to encourage consumers to pre-sort recyclable waste and deposit them in recycling "banks" (usually modified dumpsters) at regularly used locations, such as supermarket parking lots. This is currently the weakest link in the chain, because of the cost of collection—even when the consumer is inclined to cooperate—and the uneven availability of sorted product to feed the recycling systems.

However, examples of success are emerging. Esso motor oil is now being packaged in three-layer high-density polyethylene (HDPE) containers in which the middle layer is a recycled polyethylene (PE) containing 25 percent post-consumer waste. The waste comes from a reliable supply of food-grade material available in sufficient quantities to guarantee sustained production.

Early in 1989, a number of joint ventures between industry and local government were initiated in the United Kingdom. Reedpack announced a recycling joint venture for PVC and PET with the Northamptonshire County Council. Operation Recoup was launched as a nonprofit venture by thirty bottle blowers, material suppliers, and machinery manufacturers. The project covers the collection of PET, HDPE, PP, and polyvinyl chloride (PVC) bottles. In addition to its support of local initiatives, Recoup is doing research into the economics of collecting, sorting, and cleaning. It is comparing results of the Blue Box curbside recycling concept in Sheffield, inspired by a recycling program in Canada, with the European-inspired Green Bin concept in Bury and Leeds. The initial conclusions show there is a major gap between the costs of recycled plastics and the value of the recovered material. Reducing this gap will require an increase in collection efficiency linked to better cleaning and sorting methods based at optimal locations. In addition, the development of new end-use markets will help reduce the economic gap.

Another major recycling project was announced by Cabot Plastics, part of Cabot Corporation. Plans included the construction of ten recycling plants for plastics including PET and PVC throughout Europe over the next five years. Linpac Plastics announced recycling plans for polystyrene foam trays from fast food outlets. At the same time, Procter & Gamble announced it would produce plastic containers comprising 25 percent recycled material for its Lenor fabric conditioner, in a joint venture with the Newcastle County Council.

Alida announced a £3.5 million investment to recycle waste PE for the manufacture of shopping bags. This followed the success of a pilot scheme with Sainsbury's, a leading supermarket chain, to produce recycled plastic bags comprising 75 percent supermarket and post-consumer waste and 25 percent factory scrap. And in November 1989 the British Plastics Federation allocated £200,000 to a plastics reclamation project in Sheffield, which was billed as the United Kingdom's first "Recycling City."

These partnerships are beginning to work. Local government is establishing the infrastructure, consumers are becoming more aware and more responsible, and industry and retailers are responding as public pressure and the economics of environmentally safe packaging become more compatible.

Japan

Japan is one of the world's smallest countries in area, with over 120 million people living on 370,000 square kilometers (143,000 square miles) of land. It is second only to the Netherlands in population density; however, while the Netherlands is a flat country, Japan's terrain is 80 percent mountainous. With its population concentrated in the limited flat areas, Japan produces one-tenth of the world's gross national product.

In the post-World War II era, the industrialization of Japan was rapid and remarkable. Although handicapped by limited natural resources and over-population, Japanese industry concentrated on maximum productivity to catch up with the advanced industrial countries in the West. This rapid industrialization was not without its environmental consequences.

The Korean War, in the 1950s, was the first major event to trigger Japan's economic recovery. While Japan was not involved in the conflict, Japanese industry went into full gear to fulfill U.S. military orders for supplies, armaments, food, clothing, vehicles, and so forth. In the 1960s, Japan moved into the automotive industry. Economic growth meant that more people could afford to buy cars, and automotive production burgeoned to keep pace with demand. The Tokyo Olympics in 1964 was the symbolic monument to Japan's economic miracle.

In the 1970s, Japan could enjoy the same prosperity and consumerism as advanced industrial countries. Japan's material and energy consumption was accepted as a necessary part of its industrialization. If the volume of municipal solid waste is a barometer of modern civilization, then Japan had arrived. It had joined the throw-away society.

As the 1980s unfolded, the Japanese were gradually becoming aware of the importance of environmental protection. Water and air pollution, noise, and other industrial side effects were recognized as serious problems, and the public voiced growing concern over the municipal solid waste crisis. By the 1990s, global environmental protection had become the top issue. Global warming, ozone depletion, deforestation, the growth of deserts, and acid rain loomed large on the national agenda. It was no longer enough to solve domestic issues; global issues also demanded attention.

Environmental Disasters in Japan

Japan is highly sensitive to environmental issues because it has suffered from a number of ecological disasters, partly attributable to its delayed awareness of environmentalism. Its concentrated population has only exacerbated the problems. The so-called Minamata disease, caused by mercury discharged into the sea around Minamata, is one of the most notorious examples, but it was to be one of many environmental disasters in the postwar period.

Minamata disease was found for the first time in areas surrounding Minamata Bay in the Kumamoto Prefecture in 1956, and it erupted again along the Agano River in the Niigata Prefecture in 1965. The Japanese government announced that the disease had been caused by methyl mercury effluents from the plants of Chisso (Nitrogen), Ltd., and Showa Denko Company, Ltd. The mercury accumulated in fish and shellfish, which were consumed by local residents. Sufferers of Minamata disease numbered more than 1,700 by the end of 1988.

Another malady, known as Itai Itai disease, broke out in the areas along the Jintsu River of Toyama Prefecture and was first reported at an academic meeting in October 1956 as a strange disease of unknown origin. In 1968 the Ministry of Health and Welfare disclosed its view that Itai Itai disease was caused by chronic cadmium poisoning, originating from wastewater discharged from the Kamioka Mine of the Mitusi Metal Company, Ltd. The symptoms included liver and bone damage, pregnancy disorders, difficulties in nursing babies, premature aging, and calcium deficiency. As of the end of 1988, there were 124 known sufferers of Itai Itai disease.

Chronic arsenic poisoning was reported in the Toroku area of Miyazaki Prefecture July 1972, and in the Sasagadani area of Shimane Prefecture in August 1973. 160 lives were claimed in both areas.

In thirty years, Japan has suffered from a wide range of diseases and environmental disasters; a number of other disasters are listed in Figure 5-4. These problems are common to the industrialized world. While a similar chronology of oil spills, toxic poisoning and pollution could be developed for the United States and Europe, the Japanese disasters emerged within a shorter period of time and within a smaller area. These events stand as a stark warning to the industrially developing nations of the world: This is the price of ignoring the environment in the pursuit of industrial progress.

The Japanese government legislated a number of environmental regulations to address the national outcry against these disasters. Most of those are general guidelines—for example, setting maximum emission levels. Local ordinances have been more restrictive and specific. Each company works

Figure 5-4. Instances of environmental pollution in Japan.

1955	Morinaga arsenic milk poisoning.
1965	Yokkaichi zensoku (asthma).
1968	Kanemi Oil disease (caused by PCB or polychlorinated biphenyl).
1970	Photochemical smog in many cities.
1974	Mizushima oil spill into Setouchi Sea.
1978	Radioactive material released from nuclear energy plant.
1979	Nitrogen oxide and sulfur oxide became serious factors in air pollution.
1982	Acid rain.
1983	Dioxin found in wastewater of pulp plants.
1984	Tri-cloroethylene found in groundwater in large residential area.

with local government to set production controls, and the local government has the right to inspect the company periodically, with or without prior notice. As a result, the emission controls are different city by city, and in some cases vary by company.

Figure 5-5 provides a sampling of Japanese regulations.

The Green Movement

The Japanese public recognizes the seriousness of environmental issues, and activist movements have become increasingly vocal. The Japan Consumer Union, the Housewife Union, the Labor Union along with its many subgroups, and many consumer groups are among those who have been challenging corporations to change their production, R&D, and marketing strategies.

They are especially active in the following areas:

- The use of agricultural chemicals
- Dioxin effluents from paper plants
- Automobile exhausts
- The use of CFCs

Figure 5-5. Sampling of Japanese environmental legislation.

1947	Food Sanitation Law
1950	Building Standards Law
1950	Poisonous and Deleterious Substance Control Law
1951	High Pressure Gas Control Law
1957	Radioactive Isotope Control Law
1967	Basic Law for Environmental Pollution Control
1968	Air Pollution Control Law
1968	Noise Regulation Law
1969	Emission Standard for Sulfur Oxide
1970	Water Pollution Control Law
1971	Law Concerning Establishment of Pollution Prevention Organizations for Specified Factories
1971	Malodor Prevention Law
1972	Labor Safety and Sanitation Law
1973	Toxic Substance Control Law
1975	Petroleum Complex, etc., Disaster Prevention Law, Fire Service Law
1976	Vibration Regulation Law

- Noise pollution from airports and bullet trains
- Deforestation
- Phosphates in soaps
- Nuclear plant sitings

However, the political "Green" party is not yet firmly established. Some green party candidates have run for office, but none has won a seat in the Japanese Parliament.

Through the popular media, consumers are increasingly recognizing the seriousness of environmental issues. In addition, the green movements of Europe have slowly but steadily influenced the Japanese. One manifestation of this is eco-labeling, established in Japan by the Japan Environment Association, a nonprofit group, in February 1989. The group's criteria vary, but examples of products that have won the eco-label are beverage cans with stay-on tabs, magazines using recycled paper, and aerosols without CFCs. The number of designated products was 855 as of November 1990, and many industries are vying for this coveted label.

The Japan Co-op, a cooperative supermarket in Japan with 13,500 members, adopted a similar approach called the "Environmental Product System," by which certain products were selected as environmentally responsible and were promoted over competing products. The Co-op had selected 113 items as of November 1990.

Packaging

Packaging has a long and prestigious history in Japan. The ideogram for packaging contains the symbols for "care" and "gift." In Japanese society, the practice of elaborate gift wrapping is not intended to conceal so much as to adorn the gift and to communicate respect. Some of these overtones are retained in the most functional forms of packaging.

A bottle of sake wine, for example, may be sold in an ornate ceramic bottle with a handwritten label. An accompanying certificate, along with the label, is made from handmade paper. The stopper wire is chromed and polished. Finally, the bottle is wrapped in paper and packaged in a soft wooden box. This attention to detail and high quality is typical. And the affluent Japanese consumer is willing to pay for these high standards.

A spokesperson for the nonprofit Clean Japan Center was quoted as saying that the Japanese custom of overpackaging, particularly for gifts, is a "bad habit of ours." Another manifestation of this is seen in retail and grocery stores, which typically pack goods using two or three bags per item—a virtual requirement of their customers. These habits contribute to a municipal solid waste problem expected to reach 66 million tons by the mid-1990s.

Unique, multilayer, multimaterial packaging is sometimes a deliberate competitive strategy on the part of packers, aimed at preventing duplication by competitors. The increase in convenience food packaging has brought another wave of complex packaging materials, using high-performance laminates and other sophisticated materials.

From an environmental standpoint, Japanese packaging concepts may appear decidedly out of sync with green principles. However, the environmental disasters cataloged above have had a profound effect on the Japanese mind-set. While Japan is behind the United States and Europe in the application of environmental principles, the country has shown an ability to catch up quickly with Western developments, if not to exceed them.

Japan has already succeeded in developing environmentally sound packaging in certain areas, although the motivation may have included factors other than environmental consciousness, such as conservation of scarce natural resources. However, packaging design is now being shaped by environmental considerations, as shown in the examples that follow.

Glass Containers

Returnable glass bottles have a long history in Japan. The traditional container for beverages, wine, soy sauce, vinegar, oil, and other products is called Isshobin. The glass bottle was produced in a nationally standard 1,800-cubic-centimeter size that could be returned to any supplier in Japan in exchange for a deposit. While the beer bottle was slightly different in size, the same return system applied.

This uniform bottle system has continued for more than a hundred years and is still in place, saving material resources and energy while minimizing municipal solid waste. Returnable bottles are used an average of twenty times.

However, the consumerism of the 1970s and later has altered this system little by little. Consumers came to prefer the newer, better designed, and more convenient one-way bottle rather than the reusable bottle. Attractively decorated imported wine bottles, for example, took market share away from Japanese wines in traditional returnable bottles. The glass soy-sauce bottle was gradually replaced by the PET bottle because of the latter's light weight and ease of use. This consumer movement sparked a crisis in the returnable bottle system.

A nonprofit organization of winemakers has agreed to use a small, standardized, returnable bottle for wine and other beverages, as well as the traditional Isshobin.

Paper Packaging

Paper containers are used for milk, juices, beverages, and wine in Japan. They are lightweight and pose no problems for incineration—the preferred disposal method there. However, paper packaging poses environmental problems because it consumes high-quality virgin pulp and is used only once. Consumers recognize the burden it places on natural resources.

Activists have been pushing for the recycling of paper packaging, although the film coatings used to improve paper's barrier qualities create difficulties for recycling. Others advocate switching from paper to returnable glass containers. A national winemakers' association announced that it would discontinue the paper container for wine by 1993 because it is a composite structure with aluminum foil and polyethylene layers that are too difficult to recycle.

Metal Cans

Japanese steelmaking technology is very advanced; it has produced steel cans that are competitive with aluminum cans in terms of protecting the flavor of the contents. The cost of steel is also more competitive, so that the steel can is gradually recovering some of the market share that has been lost to aluminum.

Aluminum cans are used more and more for beer and other beverages because they are perceived as clean and sanitary. However, aluminum is high in cost and takes large amounts of energy to produce, which is why some beverage cans are reverting to steel. Japan's aluminum can industry is trying to increase the recycle rate for aluminum, which at this stage has reached 45 percent. However, the aluminum recycling infrastructure is not well-developed, and the results of experiments in mandatory deposit systems have not been promising.

For both steel and aluminum, the ring-pull can has long been criticized as an environmental hazard. While the United States has eliminated the ring-pull tab for beverage cans, Japanese companies have resisted switching to stay-on tabs because of the higher cost. Beverage companies feared they would lose profits and market share if they adopted the stay-on tab.

Now circumstances are changing. It is more likely today that a Japanese company will lose business if it does not adopt the stay-on tab can, because of increasing consumer awareness of environmental issues. When Suntory Company started using the stay-on tab can for its products in 1990, it was nominated as an eco-label candidate. We can therefore expect the stay-on tab to make greater inroads in Japan.

Plastic Packaging

The PET bottle became popular in Japan because it is lightweight and shatterproof. However, it has been perceived as an environmental problem for a number of reasons. PET waste requires a higher burning temperature than other waste and can damage incinerators that are not built to handle these temperatures. And PET bottles act like air balloons in landfills, making the land unstable. Some local governments are testing returnable PET bottles, but the results have been disappointing. As of 1991, there was no PET recycling in Japan.

Polystyrene foam packaging is widely used in the food industry. However, after the McDonald's hamburger chain switched from polystyrene to paper containers (see Chapter 11), other fast food retailers in Japan have adopted the same policy and are testing new paper containers. The impact of the McDonald's decision is worldwide in the fast food business.

Chlorofluorocarbons (CFCs), used as a blowing agent in polystyrene foam packaging as well as in nonpackaging applications, were recognized as destructive to the earth's protective ozone layer in an international agreement signed in 1987 in Montreal by representatives of twenty-four nations. Japan is following the schedule for reduction of CFC production that was agreed upon in the Montreal accords. CFC blowing agents are still used in polystyrene foam packaging, although they will be phased out by the year 2000. CFCs continue in use as propellants in aerosol sprays, especially in the health and beauty market, and they are also used as cleaning agents for integrated circuit boards and in refrigerants.

Most Japanese cosmetic companies, like Shiseido, have announced that they will replace CFCs with substitutes within a few years. Semiconductor makers have pledged to eliminate them by 1995, although NEC announced it will stop using CFCs by 1994. CFC makers in Japan are aggressively developing alternative substances to meet global demand. Therefore, CFC use in Japan will be drastically reduced by 2000.

Mandatory Deposit Laws

Some local governments are experimenting with a mandatory deposit system for aluminum and steel cans. These experiments have not been successful because the costs have been prohibitive. For example, a deposit of 10 yen added to the price of a 100-yen beverage poses an obstacle to buying the beverage from a vending machine, which easily accepts a 100-yen coin. In addition, the labor and freight costs associated with returning the cans is higher than the raw material costs.

On the other hand, the deposit system was successful for a long time in the returnable glass bottle market, in particular, the standard Isshobin bottle for wine and soy sauce. Although consumers began to favor the one-way bottle, with the increasing popularity of environmentalism a new deposit system may arise. There were no deposit laws for plastic containers as of 1991.

Biodegradable Containers

In order to solve the plastic waste problem, biodegradable materials are under development in major laboratories. Since Japan does not have the same requirements for airtight, watertight landfills as does the United States, degradability is seen as a meaningful concept for landfilled waste, as well as a possible solution to the problem of ocean litter.

The biotechnology division of the Ministry for International Trade and Industry (MITI) has developed a technology to control the degradable speed by adjusting the crystallization of PCL (polycaprolactone). This new material will have various applications, such as food containers and other disposable plastic items. MITI also found that PHB (polyhydroxbutyrate) is degradable without the presence of oxygen. Chuoh Chemical Company and Tokuyama Sodium Company cooperated in this R&D effort.

The largest Japanese supermarket, Daieh Company, has started using biodegradable plastic shopping bags. The cost of the bag is higher, but the company has met with a great deal of customer and employee enthusiasm for the bag. On the other hand, several environmental groups recommend that consumers bring their own reusable shopping bags to reduce reliance on the one-time shopping bag.

Paper Recycling

Paper recycling was firmly established in Japan long before the environmental issues became apparent, because Japan has limited natural resources. Government has always supported paper recycling, and all schoolchildren have been involved in gathering wastepaper. The paper recycling rate in Japan is the highest in the world. However, one popular recycling program, in which households exchanged old newspapers for toilet tissue made from recycled newspapers, has been threatened by fluctuating prices for wastepaper.

The recycling of office paper has not been so well-established. Because of concerns over the confidentiality of documents, office paper has traditionally been shredded and either incinerated or landfilled. Meanwhile, computerization has increased the volume of office wastepaper year by year.

Most offices have now begun recycling office paper. Only confidential documents are shredded. Other paper is collected in a recycling container, to be exchanged for new paper. As an additional alternative to maintain the confidentiality of data on wastepaper, watcher services are employed to monitor the movement of paper from the office to the recycling plant. In this way, even confidential documents may now be recycled as well.

In 1990, MITI announced the "Recycle 55 Plan," which would improve the recycling rate from the current 50 percent to 55 percent by 1995. MITI also announced that paper waste should be classified as industrial waste, which puts the burden on companies to pay the collection and recycling costs for wastepaper.

To save natural resources, leading companies in Japan have begun using recycled paper as much as possible, even though it costs more than virgin pulp paper. Japan pays world prices for virgin pulp because it is a world-traded commodity.

Plastic Recycling

Plastic waste is seen as a serious problem in Japan. PET bottles account for one-fourth of soft drinks sold, and demand for plastic packaging is increasing. In 1990, plastic accounted for 10 percent of municipal solid waste by weight. As much as 70 percent of this plastic waste is incinerated.

The Japanese have been slow to adopt plastic recycling, although it is now being recognized as one of the best solutions to the plastic waste problem. Some Japanese resin producers and packaging companies are investigating the recycling of certain plastics. For example, polystyrene trays are being tested for reuse as solid fuel pellets after melting them in oil at high temperatures, and other companies are working to recycle polystyrene for tray production.

Nissan Motor Company has started a trial to emboss the material name on plastic automobile parts in order to make them easy to identify for recycling after the automobiles are scrapped. Germany's BMW is experimenting with a system of "design for disassembly," and most Japanese auto manufacturers are very interested in the new approach for making cars fully recyclable.

Technology is not a critical obstacle to recycling in Japan. And Japanese consumers seem willing to participate in plastic recycling, as they have been in paper recycling. What is missing is a recycling infrastructure enjoying both industry and government support and cooperation.

Waste Collection

Curbside waste collection is very popular in Japan. Usually waste is separated into three or four major categories. In Tokyo, it is classified as combustible, noncombustible, and large-size waste. Combustible waste includes paper, rugs, organic waste, and so forth; noncombustible waste includes steel and glass; and large-size waste includes furniture and major appliances, for which a charge is levied.

The largest portion of household waste is incinerated. Nationwide there are more than 1,900 incinerators, many of which are designed to burn most plastics safely. Anything not incinerated is taken to landfills; only a small amount of municipal waste is recycled.

In local areas, waste may be sorted into many more classifications. For example, steel cans and aluminum cans are separated, steel and copper materials are separated, and glass bottles are in some cases sorted by color. However, waste separation is not nearly so prevalent as it is in Europe.

In order to minimize waste, a kind of voucher system was introduced by many local governments. Residents must purchase waste "tickets," which they exchange for waste pickup. It is an effective incentive system for reducing the total volume of waste while providing financial support for waste management.

Environmental Progress

Many leading companies have organized an environmental function during the past five years. The major missions of these new organizations include communicating with the public effectively, collecting the latest information relating to environmental issues, and both developing and enforcing corporate environmental policy. Figure 5-6 lists some of the environmental groups set up by leading Japanese companies and indicates activities and concerns with which they have been involved.

Grass-roots movements will gradually influence local corporations, which are starting to recognize the importance of working cooperatively with the local community. Because of experience gained in recycling and resource recovery, the local community is well-positioned for environmental grass-roots activity. Volunteers make use of promotional videos and cartoons to promote environmental issues. Some individuals have found money-making opportunities by participating in these activities—for example, by sorting returnable bottles or repairing broken equipment as a business.

While opportunities for partnership exist, there is still a need for greater coordination and consensus building among government, industry and consumers. MITI, which has played a powerful role in setting Japanese industrial policy, is uniquely positioned to guide research in waste management

Figure 5-6. Corporate environmental initiatives in Japan.

Shimizu Construction Company   1989
Committee for Global Environmental Issues
- Use of recycled paper
- Reduction of freon use
- Energy-saving heating system
- Prevention of desert expansion
- Reforestation

Asahi Chemical Company   1981
Headquarters for Safety and Environment
- Reduction of freon use
- Energy savings
- Waste minimization
- Use of recycled paper

Ube Kosan Company   1989
Committee for Environment and Safety
- Co-generation (producing heat and electricity)
- Waste minimization

Fuji Photo Film Company   1971
Environmental Protection Committee
- Recycling of wastewater
- Recycling of solvents
- Use of recycled paper
- Establishment of Green Fund

Hitachi Company   1990
Green Center (60 staff members)
- Development of equipment to remove nitrogen oxide and sulfur oxide from plant emissions
- Low-emission gas boiler
- Development of high-efficiency generator to reduce carbon dioxide

(continued)

Figure 5-6 (continued).

Nippon Steel   **1989**
Committee of Environment Control
Committee of Energy Savings
Committee of Global Environment
- Anti-air pollution activity
- Anti-water pollution activity
- Energy savings
- R&D for alternative products
- Promotion of tree planting on steel plant's property
- Use of recycled aluminum

NEC   **pre-1988**
Department of Environmental Management
- Development of ideal replacement for freon by 1994
- Product design for recyclability

Toyota Motor Company   **1968**
Toyota Committee of Traffic & Environment, (Staff of 16)
- Reduction of automobile air pollution
- Fuel economy
- Noise abatement
- Waste management

Tokyo Electricity Company
Environment Planning Committees at Plant Level   **1968**
Committee on Global Environmental Issues   **1990**
Waste Management Committee   **1990**
Working Group for Freon Issues   **1990**
- Reduction of air pollution in electricity generation
- Best mix of fuel for carbon dioxide reduction
- Combined cycle generator (technology for highly efficient generation of electricity)
- Supercritical plant (employs superefficient heat exchange materials)
- Superconductivity

Sony   **1990**
Packaging Committee
- Reduction of packaging material

solutions and to define business opportunities in solving the municipal solid waste crisis.

Conclusions

What accounts for Europe's highly developed environmental consciousness? The reasons may have to do with Europe's high population densities, scarce natural resources, and an early suffering of environmental degradation caused by the industrial revolution of the nineteenth century. The United States has suffered environmental catastrophes as well, but they have been spread out over a much wider geographic area. And the American public is as diverse as its vast geography; pollution of a salmon fishing ground on the West Coast may have little direct emotional impact on Americans on the East Coast or in the South.

Japan parallels many European countries in its high population density, scarce natural resources, and history of environmental disasters. However, Japan's environmental consciousness was delayed and has not permeated the political structure as deeply as in Europe. And, as we have seen, cultural and historic precedents have shaped the meaning of packaging in ways that have delayed adoption of more environmentally sound principles.

Industrially developing countries have much to learn from the examples of environmental degradation that have plagued the industrialized countries. Three forces will increasingly drive the developing countries to enact environmental management programs: international pressure from global environmental movements, domestic pressure, and pressure from multinational companies. Industrially developing nations often respond to outside pressures with feelings of nationalism, believing that no outside forces have a right to dictate domestic policy. However, the heavy price paid by industrialized countries for economic progress is a lesson in itself. Newly industrializing countries are at a stage at which prevention will prove far less costly than cure.

It should be noted that the newly democratic countries of Eastern Europe are far behind their western neighbors environmentally and face a daunting forty-five-year legacy of environmental neglect. The Elbe River flowing from eastern Germany and the Vistula River from Poland carry massive quantities of contaminants into the North and Baltic Seas. The countries of Eastern Europe offer perhaps the best argument for the positive role of public debate, adversarial challenge, and partnership among key stake-holders in the environment.

6

Tools and Guidelines for Environmental Positioning

So far, this book has focused on the current state of packaging, the packaging industry, the players who are key stakeholders in the process, and the infrastructures and technologies in place for accommodating environmental solutions. In this chapter we shift away from the conditions under which industry operates and look to the tools and guidelines that are available and upon which businesses can build their own environmental policies.

The first of these tools, life cycle analysis, is controversial and still evolving. However, the principles behind life cycle thinking are the basis of a number of emerging sets of packaging guidelines that have been carefully refined by packaging professionals, and this chapter explores some of these pioneering guidelines as well. Finally, we consider how these guidelines fit within a larger philosophical framework, which we refer to as industrial ecology.

Life Cycle Assessment

The debate on environmentally sound packaging has historically focused on packaging's post-consumer fate: the recyclability of the package, emissions from incineration, and the volume of packaging in landfills. In certain cases, hazardous substances associated with the manufacturing process, such as chlorofluorocarbons (CFCs), have also been identified. The emphasis is now shifting to the entire product life cycle, from product design and raw materials acquisition right up to the use and final disposition of the product.

Life cycle assessment is a new field of study that is being adopted by manufacturers, environmentalists, and other groups to understand the environmental positioning of products and packaging. It is a tool for analyzing environmental impacts at every stage of a package's life cycle, evaluating trade-offs, and mitigating environmental hazards. Also called cradle-to-grave or eco-balance analysis, life cycle assessment is a technique for making public policy choices as well as business decisions. Figure 6-1 illustrates the basic stages of the life cycle with which this type of analysis is concerned.

Life cycle assessment is a way of opening new perspectives and expanding the debate over environmentally sound products and packaging. However, it is not to be viewed as an end in itself. The goal of life cycle assessment is not to arrive at *the* answer, but to provide important inputs to a broader planning process.

Figure 6-1. The product and package life cycle.

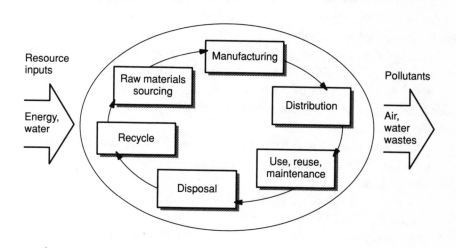

Inventories Versus Assessments

Most life cycle assessments consider raw material consumption, energy and water use, air and water pollution, and both solid and hazardous waste production. However, a meaningful life cycle assessment should be much more than a mere inventory of inputs and outputs, a balance sheet of how many toxins were produced, how much energy was consumed, or what volume of waste was generated. Along with a package's important environmental effects, life cycle assessment should also consider the overall contributions and risks to public health as well as the social, cultural, and economic impacts. The product or package should be seen in the context of the society it is intended to serve.

Historical Background

Life cycle thinking will be to the 1990s what risk assessment was to the 1980s. Risk assessment has often been used in the public policy community to develop environmental protection standards. For example, analysts used risk assessment to compare different energy sources, such as coal and nuclear power, and to study the risks of air emissions, wastes, mining, handling, and transportation associated with each energy source. The U.S. Environmental Protection Agency relied on risk assessment to establish land disposal restrictions under the Hazardous and Solid Waste Amendments of 1984 and to study chemical production and handling operations in devising the Toxic Substances Control Act, the Clean Air Act, and the Clean Water Act. In addition, the EPA's Conservation and Recovery Act is in itself a "cradle-to-grave" law, addressing the full life cycle of waste from generation to disposal. [1]

Risk assessments, which came to be embodied in the standard "environmental impact statement," are controversial procedures. The public is often disinclined to trust them, especially when conducted after the fact to justify an activity or when performed by an organization with a vested interest in their conclusions. Disputes over inputs and outputs are common. Many of these same issues persist in life cycle assessment, and they are still in the process of being resolved.

Life cycle inventories were commonly performed in the 1980s by both industry and government to compare the environmental impacts of competing products. For example, a 1984 Arthur D. Little life cycle study compared the energy usage of various containers, including glass, steel, and aluminum. The scope of the analysis was to quantify and compare the energy usage of the containers, including raw materials sourcing (mining, milling, and extracting ores), manufacturing, transportation, packing, distribution, recycling, and disposal.

Many energy uses were considered in this study. Processing methods that consume energy start with extracting and processing raw materials from the ground: melting, refining, smelting, heating, and cooling. Other energy uses considered were associated with filling the containers, from canning to bottling, including sterilization procedures. The energy costs of transportation were considered at several levels, from raw materials transport to distribution via wholesaler/retailer to waste removal facilities. The energy costs also included the cost of transportation associated with refillable containers and recycling. Costs were based on the British thermal units per ton for a given transportation mode.

All of these various energy analyses were totaled to create an energy requirement figure for each type of container. The Arthur D. Little study separately examined beer, soft drink, and food containers, as well as subcategories within those groups. Some highlights of the report:

• In fruit drink containers, the 48-ounce glass container was found to have the highest energy requirement, and the three-piece welded tinplate can the lowest, even after accounting for the greater volume of beverage delivered by the glass container.

• The two-piece steel tuna fish can was the most energy-efficient, followed by the three-piece steel can, while the aluminum two-piece can required significantly more energy.

The results of such an analysis are highly susceptible to changes in industrial practice, such as recycling rates, and Arthur D. Little has periodically revised its findings based on these changing assumptions. In addition, the Arthur D. Little study is not a comprehensive environmental life cycle assessment but one that is focused specifically on energy usage, the importance of which has moved up and down the list of business priorities depending on the price of energy. It is but one consideration in a life cycle approach that can embrace a potentially endless number of factors.

An example of a life cycle assessment was a 1990 study completed for the Council for Solid Waste Solutions (CSWS), a plastics industry task force, which compared the energy and environmental impact of paper to that of plastic grocery bags. The study found polyethylene bags superior to the unbleached paper version because the polyethylene bags produce 74–80 percent less municipal solid waste, 63–73 percent fewer air emissions, and 90 percent fewer waterborne wastes. [2] This study was limited by its assumption of zero recycling rates for both paper and plastic bags. Another study commissioned for CSWS, comparing disposable polystyrene foam packaging with paperboard packaging, used a similar methodology.

A number of corporations have developed variants of life cycle assessment for internal use. The 3M Company created a "visualization model" to help its product engineers explore the environmental impacts of new products (see Figure 6-2). While not intended as a comprehensive life cycle model, it is a tool that will be continually refined and expanded as the company seeks to establish a common foundation for environmentally sound product development. For a company that produces thousands of products and derives 25 percent of its revenues from products introduced within the last five years, this commonality will be increasingly desirable (see 3M case study in Chapter 7).

The public sector has also relied increasingly on life cycle thinking. In 1990 the EPA established a Clean Products Research Program, using life cycle assessment techniques to compare the environmental impact of products. These would be the basis for the EPA's position on an environmental labeling program to help consumers identify clean and environmentally safe products.

A program at the state level was initiated in New Jersey, in response to a task force that recommended a 60 percent recycling goal and called for criteria to phase out products and packaging not meeting that goal. The New Jersey Department of Environmental Protection began using life cycle studies to determine which products and packages were environmentally superior; these would be used to set "predisposal" fees for packages, and later for products, that generate more waste, are not recyclable or reusable, or pose environmental threats. Such fees were intended to "level the playing field"

Figure 6-2. 3M visualization model.

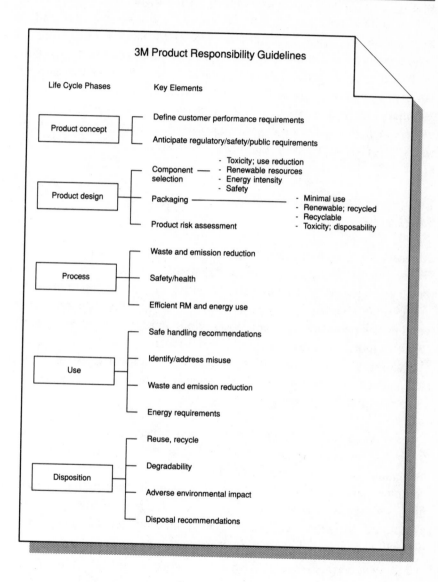

for environmentally sound products that might cost more. The California Waste Management Board also released a similar study.

The United States is not alone in embracing life cycle assessment. In Europe, eco-profiling models for life-cycle assessment of packaging have been developed in a number of countries, including Denmark, Germany, Sweden (for Tetra-Pak), Switzerland, and the United Kingdom. In Switzerland, Migros, the large Swiss retailer, has developed a software program that automates the use of life cycle data for design in order to compare the environmental qualities of alternative packages. Migros has also made the software, called Eco Base 1, commercially available.

Managing the Product Life Cycle

Manufacturers, consumer products companies, and their suppliers can use life cycle assessment as a strategic tool to study the impacts of their products or packaging at every stage in their life cycles, to seek out opportunities for competitive advantage. The possibilities include more favorable public perceptions, increased market share, cost savings, or additional revenues, while having a more positive effect on the environment.

Five areas offer opportunities for enhanced competitive position: product/package design; raw materials sourcing; manufacturing, sales, and distribution; product use; and post-consumer disposition.

1. *Product/package design.* A product and its packaging can be designed to have obvious environmental benefits. For example, leading detergent makers are developing concentrated forms of detergents, or products that combine detergent with bleach or with softeners. These solutions typically mean reduced packaging. Packaging can be designed to be easily recyclable, to be easily disassembled for recycling, or to contain no ingredients that would be harmful in landfills or incinerators.

2. *Raw materials sourcing.* Use of renewable resources and low-impact methods for extracting raw materials represent opportunities for added value from an environmental perspective.

3. *Manufacturing, sale, and distribution.* Energy efficiency, pollution prevention, and waste minimization may each offer a cost advantage and achieve a positive environmental impact. Chevron, for example, said that its SMART program (Save Money And Reduce Toxics) resulted in $4 million in savings in its first year. [3]

4. *Product use.* Competitive advantage may also stem from positioning products for an existing base of environmentally sensitive users. In Europe, marketing literature often focuses on the environmental benefits of using a product, because competitiveness is more and more dependent on environmental concerns.

5. *Final disposition.* If a collection or recycling infrastructure for a package already exists, along with a market for the secondary material, the producer may gain a competitive advantage. Recycling infrastructures are already established in some parts of the United States for high-density polyethylene (HDPE) and polyethylene terepthalate (PET), giving containers made from these materials a potential competitive advantage. Procter & Gamble has phased out its glass jar in favor of a recyclable PET jar for its Folger's instant coffee (see P&G case study in Chapter 9).

The fact remains that not everything can be subjected to a rigorous life cycle assessment. It is neither possible nor economically feasible. Common sense demands that life cycle assessments be seen as a useful tool to be balanced against other methods of analysis and inquiry.

The Communications Challenge

Life cycle studies were first performed for internal decision-making purposes, either by corporations evaluating new products and processes or by public-sector organizations evaluating policy choices. They have also been used for advertising and marketing purposes. However, for this purpose they have too often been distorted or quoted out of context to suit a particular purpose. Because of the complexity of life cycle studies, the challenge is to communicate their results in a way that promotes understanding and meaningful debate.

A danger of using life cycle assessments as a marketing tool is that they may lead to generalizations about whole classes of products, processes, or materials. For example, the study commissioned by the Council for Solid Waste Solutions, mentioned earlier, found that polystyrene foam clamshells were superior to paperboard hinged boxes, from the standpoint of energy consumption and air and water pollutants. A later study commissioned by the McDonald's restaurant chain found environmental benefits in its switch from polystyrene foam clamshell containers to paper wrap. Either study could be valid within its specified boundaries and material comparisons, but not if used to support a position about polystyrene in general.

The very fact that some packaging materials have a long and varied life cycle could be a useful concept to contribute to the public debate. Consider the example of the PET soft drink bottle. Derived from petrochemical feedstocks, plastics of all kinds consume less than 2 percent of the nation's crude oil. The PET container, the initial application of the polymer, has a limited life of a few weeks or months. In its next life, it can be recycled into a household cleanser bottle, a durable good, or a carpet fibre. Finally, the energy value of the petroleum base can be extracted when it is incinerated in a waste-to-energy facility.

In sum, the life cycle of a PET bottle shows that instead of burning and permanently depleting a fossil fuel, the plastic bottler "borrows" petroleum or natural gas to make a useful product, which can be recycled one or more times into other useful products, before the energy value is finally recovered at the end of its useful life. GE Plastics has adopted such a concept and has made a commitment to shepherding plastic throughout its many possible incarnations—in a sense, "owning" the polymer throughout its life. For example, the company is experimenting with reformulating plastics to make them more recyclable, including compatibilizers that would allow different plastics to be recycled together.

Methodological Issues

Life cycle assessment is not an exact science, and the state of the art is still evolving. A number of methodological issues remain unresolved.

First, what are the appropriate boundaries of analysis? In the example of the much-publicized cloth versus disposable diapers, do we consider the environmental implications of the manufacture of washing machines for laundering cloth diapers? Do we consider the runoffs of pesticides from cotton fields, which provide the raw material for cloth diapers? What of the possible erosion damage caused from logging operations, the first step in the paper-making process for disposable diapers? The answers to these questions are not clear.

Second, database availability is another methodological issue. If proprietary databases are used to develop a life-cycle profile, the results cannot be exposed to public scrutiny, and therefore may result in public distrust. The selection of certain databases over others may also raise objections of bias. And unless analysts rely on the same databases, assumptions, and methodologies, their life cycle assessments raise the "apples and oranges" objection.

The level of detail is a third issue. Just how many pollutants should be considered? How many manufacturers should be analyzed?

Finally, and most importantly, how do we compare results? If Product A causes more air pollution and Product B creates more water pollution, which product is more harmful to the environment? Are emissions of nitrogen oxide worse than emissions of sulfur oxide? And who makes the decision? These are thorny issues. It is the role of life cycle analysis to draw as complete a picture of environmental impacts as possible, but the final judgment rests in the hands of the industry that creates products and packaging, and the consumer who purchases them.

Some of the proposed solutions to these methodological limitations include the development of generic and publicly accessible databases, a peer review process, and standardization of methods. However, because proprietary data and methods may be seen as a part of an organization's competitive advantage, consensus building and information sharing will be hard to come by. Even if great strides are made in the science—or art—of life cycle assessment, it is hard to conclude that qualitative judgment will not always be required to achieve meaningful conclusions.

The State of the Art

In August 1990 the Society of Environmental Toxicology and Chemistry (SETAC), a 2,000-member organization with membership from industry, government, universities, and public interest groups, convened a workshop to develop a consensus on the state of the art and research needs for conducting life cycle assessments.

The workshop, in which fifty-four engineers and scientists participated, arrived at a consensus on life cycle studies that made important distinctions between three separate components:

1. *Life cycle inventory.* An objective, databased process of quantifying energy and raw material requirements, air emissions, waterborne effluents, municipal solid waste, and other environmental releases incurred throughout the life cycle of a product or process.

2. *Life cycle impact analysis.* A technical, quantitative and/or qualitative process to characterize and assess the environmental effects of the materials and energy use and environmental releases in the inventory component.

3. *Life cycle improvement analysis.* A systematic evaluation of the needs and opportunities to reduce the environmental burden associated with energy and raw materials use and environmental releases throughout the whole life cycle of a product or process.

The SETAC workshop recognized that previous life cycle assessments were too narrow in that they focused primarily on the inventory component, and its proposed life cycle concept is broader and more holistic in its approach. The group recommended a multiyear initiative to refine the inventory component, to develop the impact and improvement analysis components, to build sufficient case studies, and to communicate their use to improve products or processes.

While the authors of this book agree that life cycle assessments should be much more than mere inventories, we would go further than the SETAC recommendations and include a much broader scope of considerations. To the extent that it will be used to make substantive decisions, the object of a life cycle assessment should be seen within the context of the society in which it is used, and its impacts on the health and well-being of that society.

A product or package that fulfills a societal need must be recognized for that contribution.

In addition, life cycle assessments should recognize local or regional considerations, such as available recycling infrastructure, regulations and ordinances, climate, geographic conditions, energy and water resources, labor requirements, and other factors. Finally, such analyses need to consider personal choices, such as convenience, life-style, and cost.

Because of its systemic approach, life cycle assessment offers a more holistic and accurate perspective of the true environmental impacts of a product or package, and is a significant improvement over the single-issue focus on materials or municipal solid waste reduction. It cannot provide *the answer*, but it can offer important inputs to strategic decisions.

When the final tally of energy usage, wastes, emissions, and other quantifiable impacts are seen within the "big picture" of how the product or package is used within society, only then can judgment and reason be brought to bear, so that sound environmental choices can be made.

Packaging Guidelines

Industry must deal on a day-to-day basis with consumer demands and societal needs. Its decisions on packaging cannot wait until environmentalists, legislators, and others come to a consensus on life cycle assessment, if indeed such a consensus is possible. Nor will life cycle assessment be useful to packaging professionals in the future if it becomes the exclusive province of the scientific community, or if the outputs of life cycle assessment are not provided in useful and meaningful terms, with clear inputs to strategic decisions.

A number of organizations have already developed practical guidelines to assist packaging professionals in making the right environmental decisions based on the best current thinking. Some of the most important are described below.

IoPP Packaging Guidelines

The Institute of Packaging Professionals (IoPP) has developed landmark environmental guidelines for packaging that have been seminal to the development of packaging guidelines by leading corporations, trade organizations, and legislators. Their purpose was to help packaging professionals and corporate decision makers evaluate packaging options while assessing their impact on the environment.

IoPP, an individual membership organization with 6,500 individual members, and 150 corporate members was formed by the merger of the Society of Packaging Professionals—formerly the Society of Packaging and Handling Engineers, SPHE—and Packaging Institute International—formerly the Packaging Institute.

The genesis of IoPP's packaging guidelines was a 1988 SPHE recycling seminar. High interest in the subject resulted in the formation of an SPHE technical committee, whose first order of business was to develop a questionnaire to guide packaging professionals on environmental issues.

Procter & Gamble, Clorox, and other companies voiced support for the concept and adopted it in their packaging operations. In addition, the Coalition of Northeastern Governors (CONEG) used the concept in the development of its Preferred Packaging Guidelines (discussed later in this chapter).

In 1989 the newly formed IoPP Packaging, Reduction, Recycling, and Disposal Committee met to update the guidelines to reflect changes in the environmental challenges facing the packaging community. While acknowledging that product protection and preservation, consumer safety, and communication are the primary functions of packaging, IoPP recognized environmental impacts as a primary concern of packaging professionals.

The IoPP guidelines were designed as a sensitizing tool for all participants in the process. They also recognize the individual requirements of a company's packaging decisions; tailoring of the guidelines for company use is encouraged.

The guidelines are in question format; each package will have different functional requirements and therefore different responses. These questions are approached on three levels: theoretical, technological, and practical. In other words, a packaging material may theoretically be recycled, but there may not be a practical recycling infrastructure in place; therefore, the material would not be recyclable at this time.

The guidelines are classified into five main sections:

1. Source reduction
2. Recycling
3. Degradability
4. Disposal
5. Legislative considerations

Source reduction is the EPA's highest priority and the subject of the first set of questions. Questions such as reusability and refillability are raised. The recyclability of a package or the use of recycled content is considered the second most desirable alternative from an environmental standpoint. However, the questions recognize that recycling infrastructures are not consistent across all regions and that recycled materials must have economic practicality and a commercial recycling system through which they can be processed.

Degradability is perhaps the most misunderstood concept in the municipal solid waste field. If not exposed to the right environmental conditions, degradable materials do not break down and remain a waste product in the municipal solid waste stream. Proper exposure conditions, time frames required, and levels of breakdown must be established, and harmful by-products of degradation need to be identified. Most importantly, understanding the final disposition of the material is essential. For example, a material that degrades in water is useful if the package might be a source of marine litter, or if it is destined to be composted; it is of little consequence if the package is relegated to a modern watertight landfill.

When a package reaches the end of its useful life cycle, it must be disposed of properly. IoPP's disposal guidelines suggest that the package should be designed to facilitate its safe and easy disposal. Finally, the guidelines explore legislation that will impact packaging.

What has made these guidelines so popular? For one thing, the questions accommodate the inevitability of change: in technologies, infrastructures, regulations, markets, and consumer perceptions. Economic realities are also addressed. It benefits no one if environmentally sound practices must be withdrawn because of severe economic repercussions. In addition, the guidelines are comprehensive yet explicit. They offer guidance in identifying issues that are within the power of companies to control. They allow all participants in the process to visualize the options. Finally, they draw together procurement, product/package design, processing, marketing, distribution, and legal concerns within a company to focus on a single issue. In this sense, they are systemic in viewpoint, in the spirit of life cycle thinking.

CONEG's Preferred Packaging Guidelines

The Source Reduction Council of CONEG was established in September 1989. Recognized nationally as an innovative organization on the cutting edge of public policy development, the Council is made up of government, industry, and nonprofit groups.

The overall objective of the Council, as mandated by the nine member governors in September 1989, was to achieve "no net increase on a per capita basis in packaging waste generation over 1989 rates." The Council further recommended that this objective be expanded to "eliminate, to the maximum extent possible, packaging as a waste product as it relates to volume, weight, and toxicity." [4]

The Council's first achievement was the drafting of model toxic materials legislation that was completed and sent to the governors in December 1989. This legislation effectively prohibits the use of heavy metals in packaging four years after enactment. Specific limits were set for lead, cadmium, mercury, and hexavalent chromium in packaging sold or used within a state. By early 1991, the legislation had been passed by six CONEG states and at least four states outside the Northeast region.

The Council also established standing committees to develop a plan for implementation of Preferred Packaging Guidelines and to create a public informational program on source reduction, targeting consumers and decision-makers. The Packaging Standards Committee made source reduction the highest environmental priority for packaging, recommending the following measures in order of preference:

• Elimination of the package or packaging materials where such items are not necessary to the protection, safe handling, or function of the package contents.

- Reduction of the packaging or packing material if reduction does not result in a net increase in municipal solid waste or cause negative health, environmental, or municipal solid waste impacts.
- Design and manufacture of packaging to be returnable, refillable, and/or reusable.

Where source reduction would lead to negative impacts, the Council's alternate objective is to reduce packaging waste by, first, promoting the design and manufacture of packaging that is recyclable; second, promoting the design and manufacture of packaging containing recycled content. A package that is both recyclable and made of post-consumer content is preferred.

In short, the preferred packaging guidelines are, in order of priority:

1. No packaging
2. Minimal packaging
3. Consumable, returnable, or refillable/reusable packaging
4. Recyclable packaging and/or recycled material in packaging

As a first step in applying these guidelines, an Implementation Subcommittee developed a comprehensive questionnaire by which companies and industries could evaluate their product lines. Responses would be used to help determine opportunities for, and constraints to, packaging source reduction. Specific recommendations would be based on the questionnaire responses, expert testimony, and analyses. A Quantification Subcommittee would set definitions and timetables, while a Labeling Subcommittee would examine labels, logos, and emblems for identifying the municipal solid waste impacts of packaging.

This important work was bogged down when eleven environmental groups suspended their participation in the initiative in late 1990 and early 1991, charging that the effort was biased toward industry interests. They specifically objected to a policy adopted by the Source Reduction Council in 1990 that allowed recommendations to be adopted with a majority vote, even if conservation interests were not represented. The environmental groups were also pressing for a 50 percent reduction in packaging waste by

1995, while industry and government representatives advocated voluntary goals.

In early 1991, CONEG was working to bring the environmental groups back to the table. Meanwhile, in the spring of 1991, this council issued a challenge to the top packagers in the United States and the largest retailers in the CONEG states, asking their CEOs to formally endorse the practice of the Preferred Packaging Guidelines, to set goals for packaging source reduction, and to report goals and progress toward them to the Source Reduction Council and the public. Failing voluntary action by industry, the Preferred Packaging Guidelines could become the basis of future legislation.

Responses to the Guidelines

In addition to participating in the important work of IoPP and CONEG, industry leaders and packaging professionals are actively adopting these model guidelines in their own corporate packaging policies and practices.

One model of corporate packaging guidelines is the *Environmental Design Guidelines for IBM Packaging Engineers*, a guidebook compiled by the IBM Competence Center for Packaging Engineering. Designed by IBM engineers, designers, and suppliers with inputs from CONEG, IoPP, environmental activist groups, trade associations,and industry, the guidebook is a working tool to help packaging engineers make improvements in IBM's packaging and distribution processes, and to address and minimize the company's impacts on the environment.

IBM's design guide addresses packaging-related environmental issues from a cradle-to-grave standpoint—that is, from the purchase of raw materials and components to the handling of the finished product and parts packaging by its customers. Among IBM's goals:

- Eliminate prohibited expansion agents and heavy metals in all IBM packaging materials.
- Minimize toxic elements in packaging and in the by-products of their manufacture.
- Identify and promote the use of packaging materials that are recyclable or contain recycled content.

- Identify methods, processes, and product/package designs to reduce the municipal solid waste stream volume.
- Demonstrate that IBM is environmentally responsible and provide industry leadership in environmental issues.

To these ends, IBM set goals that met or exceeded EPA recommendations, including 50 percent recycling by 1992, and 100 percent CFC elimination by year-end 1990.

Responses to environmental packaging initiatives have come from outside the business community as well. The Institute of Food Technologists (IFT), which represents the scientific community, has expressed concern that the emphasis on post-consumer waste issues may obscure the basic function of packaging.

IFT conducted a workshop in November 1990 on "Food Packaging, Food Protection and the Environment." The group expressed concern that environmental legislation or regulation addressed solely to after-use disposal of packaging may actually jeopardize public health. IFT emphasizes that the primary function of food packaging is safety: protection from spoilage, poisoning, infestation, and disease. These goals are not served by divergent packaging standards from state to state.

In the interest of ensuring food safety, IFT scientists declared the following as part of an extensive list of recommendations:

- Insist that safety, quality and nutrition remain the primary requisite for food packaging.

- Oppose bans and taxes on specific packages or packaging materials for the sake of environmental policy unless impacts on food safety, quality, and nutrition are assessed.

- Support integrated waste management methods, including source reduction, composting, recycling, waste-to-energy conversion, anaerobic digestion, landfilling, and new technologies, as an integral part of local and/or regional municipal solid waste management systems.

- Encourage source reduction as the preferred means of reducing municipal solid waste.

Industrial Ecology

As these examples indicate, there is now a broad social recognition that the environment is a serious issue. Significant systemic change is unavoidable if society is to address the root causes of environmental damage. Effective policy and strategy need to be based on a farsighted conceptual framework.

Industry has recognized and is already acting on classic environmental problems, such as toxic pollutants. Now the scale of industrial production is such that even nontoxic emissions, like carbon dioxide, have become a serious threat to the global ecosystem. The key issues are systemic, rather than localized. The current focus on life cycle assessment is indicative of the recognition that a systemic approach is needed.

The typical environmental business posture of yesterday was reactive, with companies acting largely on the impetus of legislative pressure or public opinion. In their desire to remain profitable, companies favored solutions that pandered to common perceptions and usually those involving the minimum effort required to ensure compliance and end-of-pipeline cleanup. The risks involved in this posture included unforeseen costs and acceptance of a follower role.

The emerging "green" corporation of tomorrow accepts the environmental imperative and willingly assumes the mantle of environmental leadership. It favors voluntary product and process redesign, as well as the avoidance of pollution and waste. In short, it takes the long-term view and addresses environmental issues by attacking their root causes.

However, industry still lacks a coherent conceptual model of its overall environmental objective. Such a model should be forward-looking and entrepreneurial, with a bias toward innovation and action. It should also be compatible with business activity and contribute to the company's competitiveness. Most importantly, such a model should recognize that no one corporation acts alone; all companies and all industries are part of a larger interlocking system in which each part affects the whole.

Such a conceptual model exists. Patterns of industrial activity can be modeled on patterns of biological activity, leading to the concept of an "industrial ecosystem." This paradigm sees the natural environment—traditionally the focus of the problem—as the model for the solution.

The most important feature of the natural global ecosystem is that all outputs are inputs somewhere else. There is no such thing as "waste" in the sense of something without use or value. Materials and energy are continually circulated and transformed. From the death and decay of one species come the life-giving nutrients for another. Thus, all parts of the system act independently, but are meshed cooperatively.

These are some of the important features of an industrial ecosystem:

- Compatibility with the natural system
- Maximum internal reuse of materials and energy
- Selection of processes with reusable waste
- Extensive interconnection among companies and industries
- Sustainable rates of natural resource use
- Waste intensity matched to natural process cycle capacity

Human modification and manipulation of ecosystems is as old as agriculture. We now need to integrate industry into the equation. A fully developed industrial ecosystem will emerge as ecologically-based technology progressively replaces the existing industrial infrastructure (see Figure 6-3).

Figure 6-3. Timeline of evolving corporate environmental response.

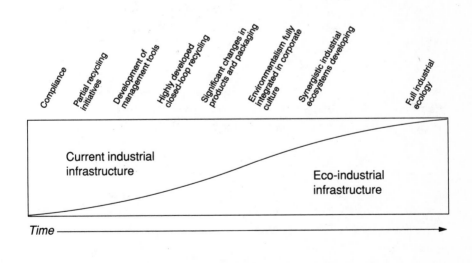

165

While the idea of industrial ecology may appear idealistic or even far-fetched, it is in fact the most plausible model for sustainable industrial development. Researchers at leading-edge organizations such as Bell Laboratories, Carnegie-Mellon University, and Princeton University's Center for Energy and Environmental Studies are actively studying the concept. In addition, major corporations are in effect already putting the industrial ecology idea into practice. The case studies that follow this chapter are early indicators of an emerging industrial ecosystem:

• 3M Company, with one of the oldest and most successful environmental programs, stresses pollution and waste prevention, attacking the problem at the source.

• Du Pont, one of the world's largest chemical and plastic producers, is voluntarily eliminating CFCs and has found ways to convert both industrial and consumer waste into commercial products.

• Procter & Gamble, which has applied environmental thinking to some of the world's best known consumer products, has worked to build a recycling infrastructure by creating linkages with other companies up and down the supply chain.

• Johnson & Johnson has found ways to link its worldwide system of companies in a common response to the complex issues of the global environmental condition.

• The National Polystyrene Recycling Company offers further lessons on the importance of the interconnections between the many players in the system.

Although these companies may not explicitly recognize the concept of industrial ecology, the early signs of an industrial ecosystem are evident in their policies and practices, designed to have a minimum threat to the environment while supporting sound business practices.

Notes

1. Examples cited in Paul Bailey, "Life Cycle Costing and Pollution Prevention," *Pollution Prevention Review*, Winter 1990-1991, pp. 34–35.
2. Franklin Associates, "Resource and Environmental Profile Analysis of Polyethylene and Unbleached Paper Grocery Sacks," 1990.
3. This and other examples of competitive advantage were drawn from Karen Blumenfeld, "Managing the Product Life Cycle," *Management Review*, March 1991, pp. 30–31.
4. "Source Reduction Council of CONEG: Progress Report," March 29, 1990, p. 3.

7

Pollution Prevention: 3M Company's 3P Program

Minnesota Mining and Manufacturing Company, better known as 3M, has perhaps the longest running and most successful environmental program in U.S. corporate history: the Pollution Prevention Pays or 3P Program. While this program does not directly focus on packaging and municipal solid waste issues, it is a valuable case study of how a sound environmental policy and a concrete set of programs can become institutionalized in the corporate culture.

Initiated in 1975, the 3P Program was not an isolated case of environmental goodwill, but a permanent and continuous program of action that has resulted in more than 2,700 successful projects in its first fifteen years, while yielding $500 million in savings for the company and a 50 percent reduction in pollution per unit of production.

This is also a story of how 3M's corporate culture may be uniquely well-suited to enacting environmental policy in a deep and comprehensive way.

Background

Founded in 1902, 3M is a complex and diverse multinational company with more than $13 billion in sales worldwide as of 1990. It has forty-five major product lines and thousands of individual products, from industrial adhesives and sandpapers to surgical supplies and recording tapes, as well as the ubiquitous Scotch brand tape and Post-it brand notes. 3M's divisions are grouped along technology lines in four major businesses: information and imaging technologies, commercial and consumer, industrial and electronics, and life sciences.

3M is a technology-driven company. Unlike a conglomerate that grows by acquiring unrelated businesses, 3M grows its own businesses by improving and expanding its technology base. The company's top managers tend to be engineers, as opposed to accountants, salespeople, or attorneys—a further reflection of the technological underpinnings of the company. It is also a company that promotes from within, and it is typical for a 3M professional to devote a lifetime career to the company. As one 3M manager put it, "A vice-president at 3M with less than twenty-five years of experience with the company would be on the fast track. You won't see any thirty-year-old wonders in top executive offices."

3M's credo of growth from within has particular ramifications for environmental policy. It reinforces the long-term outlook that is necessary for meaningful environmental improvements. In practical terms, it means that a plant manager who oversees an area with a pollution problem may see that same problem come home to roost years later, when he is further up the corporate ladder—and that manager will be held accountable for it.

In addition, 3M's legendary ability to maintain an entrepreneurial culture is conducive to innovation in many forms, including environmental improvements. This culture is based partly on corporate structure, in which the division is the basic business unit of 3M. Each of the roughly fifty divisions is a fairly complete business in itself, responsible for its own research, manufacturing, and marketing. The divisions are flexible and free to move quickly to seize business opportunities, while being backed up by the technical and distribution strengths of the company.

A less tangible part of the corporate culture is 3M's ability to inspire teamwork and cooperation. 3M people are remarkably attuned to networking and sharing information with others in the company. 3M encourages this cooperative spirit by recognizing innovation, but it does not provide monetary rewards for innovation at the individual level. The company believes that monetary incentives encourage unproductive competition, secrecy, and holding back information. Through top management recognition and opportunities for advancement, 3M fosters openness and participation.

The divisions compete within the sectors for corporate funding, based on their performance, and this provides an additional incentive for developing new products. One of 3M's goals is to derive 25 percent of sales from products introduced within the past five years. 3M wants to be the one to make its own products obsolete. The result is a steady stream of innovations.

The company also offers dual career ladders; it is possible to be promoted up a management ladder as well as up a scientific ladder. There are eight levels in the scientific hierarchy, with the corporate scientist position equivalent to a vice-presidency. In this way, scientists do not have to give up science to advance within the company. This policy allows individuals to continue to do what they do best throughout their careers. If a talented individual fails at a certain job, it is assumed that the job was wrong, not the person—and the employee is shifted to another position.

3M encourages its employees to spend 15 percent of their time on projects they personally consider worthwhile, and environmental projects fall within this category. Individuals with strong track records are allowed much greater flexibility. And some corporate scientists are allowed almost unlimited freedom to champion their projects.

Genesis of the 3P Program

The early 1970s saw a proliferation of federal environmental legislation, including the Clean Air Act, the Clean Water Act, and the formation of the Environmental Protection Agency. The decade was also a period of growing environmental activism. Although the EPA had yet to impose any major regulations, 3M recognized a rising crescendo of environmental awareness that would ultimately bring hefty compliance costs. By 1974, the United States was also moving into a recession, and upper management was looking for ways to cut costs.

An example of the company's pollution dilemmas may be seen in its use of solvents. Many of 3M's products are made with coating processes. Typically, coatings are dissolved in solvents, so that they may be applied evenly and thinly; the solvents are then dried off with heat. The problem is that, as they dry, solvents like toluene, xylene, and methyl ethyl ketone are released into the air, where they may combine with nitrous oxides and sunlight to create smog. Pollution control equipment can reduce as much as 85 percent of these air emissions; the captured solvents may either be incinerated or recovered for reuse.

Regulations *were* being met through end-of-pipe pollution controls, but these "add-ons" were expensive to operate and still allowed 10 to 15 percent of solvents to be emitted into the air. While better than nothing, pollution controls are not as good as finding ways to prevent the problem in the first place. 3M began looking for lower cost and longer lasting solutions.

The combination of environmental and cost concerns grew into an official 3M environmental policy, which took 3M from a reactive position to one of positive action and self control. Under the policy, 3M is committed to do the following:

- Solve its own environmental pollution and conservation problems.
- Prevent pollution at the source whenever and wherever possible.
- Develop products that will have a minimum effect on the environment.
- Conserve natural resources through the use of reclamation and other appropriate methods.
- Ensure that facilities and products meet and sustain the regulations of all federal, state, and local environmental agencies.
- Wherever possible, assist government agencies and other official organizations engaged in environmental activities.

Following development of this policy, new programs were designed to carry the policy into action. The 3P program was one of them. Recycling, commuter-ride sharing, and energy conservation are among 3P's companion programs within the company.

The idea behind 3P was simple: Prevent pollution at the source in both products and manufacturing processes, rather than remove it after it has been created. Although the idea itself was not new, the idea of applying pollution prevention on a companywide basis around the world, and recording the results, had not been done before 3M's initiative.

The story goes that Dr. Joseph Ling, who then headed 3M's Environmental Engineering & Pollution Control (EE&PC) activity, said, "We have two choices. First, we might choose not to comply with regulation, and then we figure out which of us goes to jail. Second, we choose not to create pollution in the first place." This comment was said to have been the inspiration for the 3P Program. In reality, Dr. Ling and others who initiated the program had a background in wastewater treatment, and this training taught them always to go back to the source of a problem. They had also been working with water-based adhesives to address pollutants from solvents.

Dr. Ling's environmental engineering group consisted of twenty-five people at the time. It was a corporate group, not an operating unit, which gave 3P's proponents immediate access to top management. After winning the enthusiastic approval of then-CEO Raymond Herzog and the board of directors, the 3P Program was formally introduced throughout the company in 1975.

The company needed to market the program internally, but not with a big, splashy introduction that might be quickly forgotten. The company chose a continuous, low-key approach to sell the idea, to infuse it permanently in the corporate mind-set. The idea was to get people to think about pollution prevention in the course of doing their jobs every day.

A combination of videos, slide shows, seminars, written program materials, and award programs was designed and assembled to market the program to employees. A key element was the development of a symbol, a "little green man," which was the subject of a series of cartoons posted on internal bulletin boards. Every week, a new cartoon would appear, with the little green man depicting some new idea in pollution prevention. The green man was a simple, but attention-getting, symbol, and it became the logo of the program. The cartoon series was a major part of a corporate motivational program so successful that it won a Silver Anvil award from the Public Relations Society of America.

Initially, the 3P Program was not actively marketed outside the company. Before 3P had established a track record, the company's public relations people were concerned that a public announcement would invite a backlash. Today, the program is so widely recognized that 3M receives many unsolicited requests for information. 3P has been copied by a number of companies and has received numerous environmental achievement awards.

The 3P Program had an outside publicity boost in 1977 when the federal EPA and the Department of Commerce cosponsored four regional conferences on pollution prevention. The basic concept behind these high-level conferences was that pollution prevention is the best way to address environmental goals. Unfortunately, when the Clean Air Act amendments came out later in 1977, the regulatory focus shifted to pollution cleanup rather than prevention. Nevertheless, 3M stayed on the prevention course, and it was not until the late 1980s that prevention would be taken up again in the legislative structure.

The 3P Program

3P encourages technical innovation to prevent pollution at the source through four methods: product reformulation, process modification, equipment redesign, and resource recovery. Projects that use one of these methods to eliminate or reduce pollution, save resources and money, and advance technology or engineering practice are eligible for recognition under 3P.

3M's vice-president for environmental engineering and pollution control is responsible for implementing 3P within the company. Environmental engineers within EE&PC are assigned to particular operating divisions to ensure compliance with all government and 3M standards, and implementing 3P within those divisions is a part of their responsibility.

Typically, 3P programs are initiated when an employee recognizes a specific pollution or waste problem and a possible solution. A cross-functional team is developed to analyze the problem and develop solutions; this team might be composed of employees from several divisions, such as engineering, research, marketing, and legal. Using a simple submittal form, the team submits a proposal to the affected operating division, and a decision is made on whether to commit funds, time, and resources.

Here are some examples of approved 3P projects:

- A 3M facility in Alabama recycled cooling water that previously had been collected for disposal with wastewater. Reusing the water allowed 3M to scale down the capacity of a planned wastewater treatment facility from 2,100 gallons a minute to 1,000 gallons a minute. The recycling facility cost $480,000, but 3M saved $800,000 on the construction cost of the waste-water treatment plant.

- A resin spray booth had been producing some 500,000 pounds of over-spray a year that required special incineration disposal. The booth was redesigned and new equipment was installed to eliminate excessive over-spray, resulting in a net reduction in the amount of resin used. The savings amounted to $125,000 annually on a $45,000 investment. [1]

In the course of fifteen years, the EE&PC staff has increased from twenty-five to eighty people, and its record of achievements has grown as well. Worldwide annual releases of air, water, sludge, and solid waste pollutants from 3M operations has been reduced by half a million tons, with about 95 percent of the reductions coming from U.S. operations.

One of the reasons behind 3P's success is the company's belief that the people who do a particular job are the experts. They are the best qualified to know how to solve problems that originate in their areas. As Thomas Zosel, manager of Pollution Prevention Programs within EE&PC, explained it:

Environmental engineers are usually trained in providing end-of-pipe solutions. I may know how to install a thermal oxidizer, but I can't teach people at the divisional level what to do with their processes. I don't have the expertise or background. We can point out the problem, but we can't solve it for them. The divisional people need to do the actual work.

The 3P Program achieved the goal that was set forth by former chairman Herzog "to continue indefinitely until it was ingrained in the corporate culture." Remarkably, the program continued in force with no change for its first thirteen years of operation. However, while the basic concept of pollution prevention remains timeless, 3M management concluded in 1988 that 3P's results did not go far enough. Emission reductions were significant, but the company wanted to do more. Therefore, a short-term program called Pollution Prevention Plus (3P+), was initiated.

3P+ Program

As 3M CEO Allen Jacobson noted, "With 3P+, we are moving into new territory, taking environmental responsibility well beyond what is required. With the idea of prevention thus ingrained in our operating philosophy, we move a step closer each day to the ideal of zero pollution." [2]

At the heart of 3P+ is a long-term research effort to reduce sources of pollution in their manufacturing processes—the classic pollution-prevention approach. The goal is to cut all releases to the environment by 90 percent from 1987 levels by the year 2000.

The 3P+ Program is a more structured effort than the voluntary 3P Program. Waste minimization teams are being formally established in every 3M operating division to identify source reduction and recycling opportunities and to develop plans to address them.

As an intermediate step, to help 3M reach its new goals quickly, the company is investing $170 million in air pollution control equipment to reduce annual hydrocarbon emission by 55,000 tons—over and above other reductions achieved through pollution prevention. The goal here is to cut air emissions by 70 percent from 1987 levels by 1993. These are not arbitrary figures, but projections based on specific and detailed projects.

Between 1988 and 1992, 3M is installing additional control equipment at all plants emitting more than 100 tons of hydrocarbons annually, even if they meet all pertaining air quality standards. Facilities outside the United States are included in the plan, although their time frame is extended to mid-1993.

One important feature of 3P+ is to donate emission credits to local permitting agencies, rather than sell them for its own profit. Under current environmental regulations, companies that make emission reductions in certain parts of the country can earn credits, which they may sell to other companies that may then increase their emissions. Environmentalists have criticized the credit system as a "license to pollute." 3M will not sell its credits for its own profit, although it has considered selling the credits and turning the money over to a local county for environmental programs, if the net result is an environmental improvement. 3M retains only those credits needed for its future plant needs, such as equipment purchases, and donates the rest to regulating agencies. For example, in 1987, 3M returned 1,000 tons of air emission credits to the state of New Jersey, and in 1988 it donated 150 tons of air emission credits to the Los Angeles area. These Los Angeles credits alone could have been sold for more than $1 million. 3M takes a more modest tax deduction instead.

In addition, 3P+ marks a change in direction from 3P in the way projects are cost-justified. *With or without cost savings*, 3M would do what was necessary to protect the environment. While formerly the 3P projects had to meet the same return-on-investment criteria as any other project in the company, the new 3P+ policy takes into consideration more intangible benefits, like the enthusiasm and morale of its employees, and the desire to do the right thing for the environment.

3M is betting that environmental protection will result in better products, lower costs, and a stronger competitive position. However, this is based on intuitive judgment, a leap of faith, rather than strictly financial criteria. Even the $170 million investment in hydrocarbon emission controls was made without any expectation of return on investment. As EE&PC Staff Vice-President Dr. Robert Bringer states it, "Over the next ten years we're converting to products that don't pollute. There may not be an obvious return today, but ten years from now we'll be in a much better competitive position."

3M management believes that by getting ahead of the game, it will not be forced into a crisis mode when stricter regulations are enacted. Rather than having to implement quick-fix solutions that may not be technically or economically sound, the company is buying time to study the issues and choose well-engineered solutions that provide the best long-term benefits.

And by having proactive programs, the company can set the agenda for environmentalism, rather than be put on the defensive by environmental activists. Few environmentalists would disagree that prevention is the best solution to pollution problems. 3M is one of fifteen companies that are active in the Corporate Conservation Council within the National Wildlife Federation, and 3M communicates as circumstances dictate with the Environmental Defense Fund and other environmentalist groups.

Beyond 3P

In 1990, 3M's environmental engineering group presented two new goals to the corporate operations committee, neither of which could be quantified: to introduce products that had lower environmental impacts and to apply pressures on company suppliers to do the same. In this way, 3M is responding to requests from customers who want to know what the company is doing to make its products greener, while raising the same questions with its suppliers.

Some customers are already asking 3M to take back waste from its products. That in itself is an incentive to reduce waste. 3M has also worked with suppliers to replace hazardous chemicals and to develop substitutes. For example, 3M is well under way toward achieving its goal of eliminating stratospheric ozone-depleting chemicals: chlorofluorocarbons (CFCs) by the end of 1990, carbon tetrachloride by 1991, and methyl chloroform by 1992.

The company has also begun to recognize products and packaging within a "beyond 3P" Program, rather than just processes. 3M strives for products that offer environmental improvements, but it shuns the term "environmentally friendly." 3M considers the phrase meaningless and misleading, because virtually anything can have a negative environmental impact. The company takes a cautious stance on green product claims, and a committee made up of marketing and environmental people reviews all such claims before they go public.

For example, 3M changed the formula of its Scotchgard fabric protector to eliminate CFCs and trichloroethyelene. However, the product still contains volatile organic compounds that could contribute to smog. Therefore, the label now indicates that the product contains no ozone-depleting chemicals, but it does not claim that the product makes an environmental improvement.

If a product is made from recycled material, 3M indicates the percentage of recycled content. The company is incorporating recycled fiber in some of the most popular sizes of its Post-it brand note pads. The pads use a minimum of 50 percent recycled content, or at least 10 percent post-consumer waste by weight.

In the packaging area, the company has found a competitive edge in reusable packing materials. 3M is one of the few remaining U.S. manufacturers of videotape. Its Hutchinson, Minnesota, plant sells bulk videotape to companies that duplicate movies for sale to the public. The large rolls of film are shipped in protective foam packaging, which is collected at customer sites and returned to 3M for reuse again and again. This saves the customer from having to pay to dispose of the materials and reduces impact on landfills. As the manager of 3M's videotape plant indicates, "This gives us an edge over suppliers from Japan or Korea. They are in no position to recycle their packaging."[3]

Another of 3M's product-related initiatives is to apply the principles of life cycle analysis to all its major product lines. For this purpose, it has developed a one-page "visualization model." Not intended as a comprehensive checklist or strict model for life cycle analysis, it is designed as a sensitization tool (see Chapter 6 for depiction of the model). As the operating units develop new products, the model will help 3M scientists analyze new product concepts from environmental, health, and safety perspectives, as well as the traditional performance and quality criteria, and to anticipate environmental issues from design to disposal.

Life cycle thinking has also extended to the company's waste management philosophy. As 3M's environmental consciousness has evolved, it has identified four new goals underlying its corporate waste strategy:

1. *To do the environmentally positive thing.* Doing what should be done, rather than what has to be done, means that economic objectives should be broadened to include intangible benefits like corporate reputation and employee morale.
2. *To reduce current and future liabilities.*
3. *To pursue cost-effective solutions.* No waste management system will work unless it makes economic sense to do so.
4. *To improve continually the quality of product and processes.*

Together, these four aims and 3M's environmental policy have evolved into its current program for managing waste from "cradle to grave." The hierarchy of waste management has shifted away from pollution control to prevention of waste generation, followed by reclamation and recycling, treatment, and proper disposal of residuals. 3P, with its emphasis on reducing hazardous and nonhazardous waste in land, air, and water, remains an integral part of this approach.

3M believes that competitive forces will drive environmental progress. Most of its customers will not sacrifice quality, performance, and price for the sake of environmental improvements. However, if its products compete on all those criteria and are also "greener," 3M will have a competitive edge. Environmental considerations can also help 3M become a lower cost producer.

Notes

1. Examples cited in Thomas Zosel, "How 3M Makes Pollution Prevention Pay Big Dividends," *Pollution Prevention Review*, Winter 1990-1991, pp. 69–70.
2. Quoted in Dr. Robert Bringer, "Pollution Prevention Plus," *Pollution Engineering Magazine*, October 1988.
3. John Holusha, "Hutchinson No Longer Holds Its Nose," *New York Times*, February 3, 1991, Section 3, p. 6.

8

A Leadership Role: Du Pont's Environmental Policy

Du Pont has long been recognized for its environmental leadership. Its executive committee made its first commitment to environmental stewardship in the 1930s, and since then it has applied tough environmental safety standards on a global basis.

The company's environmental leadership found a powerful new impetus in the spring 1989 appointment of Edgar S. Woolard, Jr., as chairman. From the beginning, Woolard made corporate environmentalism one of Du Pont's key business responsibilities. This was much more than a token effort to be a "green" company. It was a top-down executive mandate to make environmentalism as important as financial success and safety performance. All senior managers would be measured in terms of how well they managed environmental responsibility—and it would be reflected in their paychecks.

Du Pont is a multinational corporation, with 45 percent of its business outside the United States, and its environmental perspective is global as well. From the beginning, Woolard personally delivered his corporate environmentalism theme in speeches throughout Europe, Asia, and the United States. It was a fundamental new business direction based on waste minimization, leadership in plastics recycling, community involvement in environmental issues, and wildlife habitat enhancement. The company even launched an environmental business, called Safety and Environmental Resources, to serve its internal businesses as well as to provide company customers with environmental solutions.

Within recent years Du Pont's commitment has been recognized outside the company as well: In a 1990 survey by *U.S. News and World Report*, 40 percent of American business leaders named Du Pont as the most environmentally responsible company.

Plastics Initiatives

Among its many business interests, the Delaware-based chemical giant is one of the largest and most diversified producers of plastics in the world, with a major market in packaging. The company has introduced more than 30,000 plastic products.

An important focus of its polymer business is plastics recycling, first in industrial plastics, more recently in the recovery of post-consumer plastics. To oversee these efforts, the business established a management position, that of director of environmental affairs, responsible for focusing on the plastics waste management issue worldwide. Frank Aronhalt, the first to hold this position, was also given business and operational responsibility for all Du Pont's plastics recycling operations. This marked the first time a major resins producer combined issues-management responsibility in the environmental area with an actual profit-and-loss responsibility.

Du Pont had already been involved in industrial plastics recycling for more than a decade before it moved into post-consumer recycling. The company has twelve plants for industrial plastics recycling, of which eleven are in the United States and one is in the Netherlands; and the director of environmental affairs has operational responsibility for several of these facilities as well. These plants do more than in-plant recycling, that is, returning industrial waste to primary manufacturing processes—a fairly common practice. The facilities process byproducts from Du Pont's manufacturing processes, as well as from its customers' waste stream, and turn them into entirely different products.

An example is Du Pont's Cronar X-ray film. One of the company's plants buys back the used film from hospitals and recovers the silver and the polyester film base. Anyone who has ever flown on a domestic airline will have seen the endproduct of recycling Cronar and similar films: the clear plastic covers of airline magazines made from recycled Cronar. This and many other businesses in industrial recycling, taken together, process 200 million pounds of plastic per year, generating more than $55 million in annual sales. And this is entirely separate from Du Pont's more recent post-consumer recycling initiatives.

However, few people would have known about these industrial recycling initiatives, even as recently as the late 1980s. Because Du Pont derived a competitive advantage from such facilities, the company was reluctant to publicize its successes in this area. These industrial recycling plants were viable, profit-making businesses, and Du Pont kept them close to the vest. Now, however, recognizing the need to tell its recycling story, the company is no longer quiet about its industrial recycling achievements.

Building on its industrial recycling successes, Du Pont moved into post-consumer plastics in 1989 with the Plastic Recycling Alliance, a joint venture with Waste Management, Inc. (WMI). Aronhalt has been involved with the joint venture since its beginning and was part of the negotiating team that created it. The Plastic Recycling Alliance plans to establish five recycling plants in North America dedicated to certain post-consumer plastics, including polyethylene terephthalate (PET), high-density polyethylene (HDPE), and multilayer constructions. This alliance was logical; it brought together the largest waste management company in the world with one of the world's largest and most diversified producer of plastics. It combined WMI's experience in collecting and sorting post-consumer plastics and Du Pont's expertise in plastic technology, recycling, and market access. And it demonstrated both players' belief that plastics recycling would never be a reality unless collection was linked with markets.

The joint venture would establish one of the country's largest and most comprehensive plastic waste management systems, and it represented an important investment in the recycling infrastructure. The first two plants, in Philadelphia and Chicago, were up and running in 1990, each with a capacity to recycle about 40 million pounds a year. The full North American network, to be completed by 1995, was expected to have the capacity to recycle some 200 million pounds of post-consumer plastic annually.

The Plastic Recycling Alliance is one of several Du Pont initiatives to help solve the plastic waste dilemma. The following solutions are representative:

• *Internal waste minimization.* Du Pont has formal waste minimization programs at each of its worldwide manufacturing sites. These basic polymer production facilities recover and reuse nearly 1 billion pounds of in-plant polymer wastes and polymer intermediates annually, without compromising finished product quality.

• *Packaging awards.* As a part of its prestigious annual food packaging award program, Du Pont established an environmental award category, designed to recognize innovations that reduce the amount of plastic packaging in the municipal solid waste stream. Procter & Gamble's Spic and Span Pine bottle, made from 100 percent recycled PET, was the category's first winner in 1990. Judging is done by a distinguished panel selected from the packaging industry, government, and academia.

• *Employee programs.* Some 20,000 Du Pont employees drop off used PET and HDPE bottles at Wilmington, Delaware, collection sites; Du Pont makes contributions to employee-designated charities for each pound collected. There are also corporate programs for recycling office paper and aluminum beverage cans.

• *Recycling pesticide bottles.* In 1989 Du Pont began cosponsoring a pilot program with the National Agricultural Chemicals Association, the state of Mississippi, Washington County, and other agricultural product manufacturers to study the feasibility of recycling HDPE pesticide containers into new agricultural product containers, as a safe alternative to incinerating or landfilling. Initial results were encouraging, and other pilot projects were held in Minnesota, Florida, and Iowa. As a result of this program, Du Pont introduced two of its herbicides packaged in recycled plastic jugs.

• *Beach clean-up.* Du Pont worked with communities in Delaware, Texas, and Florida to collect thousands of pounds of plastic waste from beaches and recycle it into park benches. These benches were returned to the communities, giving useful life to what was once beach litter.

• *Food service recycling.* The cafeteria at Du Pont's Camden plant leases a machine from Dart Industries, a major producer of polystyrene products, that compacts food trays and other polystyrene materials on-site into

40-pound logs that Dart picks up for recycling. This on-site processing eliminates the objection that shipping foam is too costly, because of the material's high volume and low weight. The plant managers can continue using food service products they prefer, while saving money over the alternatives, such as washing dinnerware.

• *Motor oil bottle recycling.* Conoco, Du Pont's petroleum operation, has developed a program in Germany to collect used polyethylene motor oil containers at gas stations, where they are compressed and returned for recycling into oil bottles with recycled content.

• *Automobile components.* The company has established programs in the United States, Japan, and Europe to develop expertise and economic capability for recycling major plastic components of automobiles.

These initiatives are consistent with the company's total environmental stewardship philosophy. "You can't just create a product that does a wonderful job," Aronhalt said. "We now need to think ahead to how the product is disposed of when it has finished its useful life. To do anything less is to risk product failure."

Top-Down Empowerment

On May 4, 1989, one week after beginning his tenure as Du Pont's chairman, Edgar Woolard announced at a meeting of the American Chamber of Commerce in London that he was not only Du Pont's chief executive, he was also its chief environmentalist.

"Our continued existence as a leading manufacturer requires that we excel in environmental performance and that we enjoy the nonobjection—indeed even the support—of the people and governments in the societies where we operate around the world," Woolard told the London gathering.

However, how does one translate a speech into actions? Corporate management needed to be motivated to get the job done. Woolard made it clear that a part of all senior executives' compensation would be based on how well they managed their environmental responsibilities. It was analogous to the company's long-standing practice of rewarding employees for safety performance. In practical terms, it meant, for example, that if a plant manager had an environmental mishap at the plant site that could have been prevented by good management practice, the manager would feel the consequences financially. Salary and incentive compensation were directly tied to environmental performance.

Woolard let it be known what he expected from Du Pont employees. For one thing, any new business plan had to include a statement of how the manager planned to address environmental responsibility by actions, by programs, or other means. That requirement became especially relevant in the packaging area, where waste minimization, recycling, and recovery are well-established goals. Woolard also communicated the point that each one of its manufacturing facilities exists at the pleasure of the local community. This is a practical recognition that if the plant does not meet community standards, it will not survive in the long term. The company's philosophy has encouraged a new openness with civic and environmental leaders.

While everyone at Du Pont is expected to be cognizant of environmental issues, there is a structured organization of professionals devoted full-time to environmental affairs. Du Pont's plastics business alone has nearly fifty people worldwide working on the plastic waste issue, as distinct from hazardous waste management or plant operations seeking practical solutions and new business opportunities. The director of environmental affairs for polymers devotes about one-third of his time to business management and two-thirds to issues management. His organization is responsible for recycling operations, market development of uses for recycled plastics, research and development in such areas as recycling, degradable polymers, and removal of heavy metals content. A corporate communications and government relations staff are dedicated to environmental issues.

A counterpart responsibility exists in Canada, Europe, Japan, and Australia. In addition, a companywide committee brings together over forty people from every Du Pont and Conoco business worldwide that is either currently or potentially impacted by the plastics waste issue: packaging, textile fibers, automotive, agricultural products, and so on. These staffers help set program directions and identify business opportunities within the context of environmental policy.

The CFC Story

Perhaps the most dramatic example of Du Pont's environmental policy was the decision to discontinue its $750 million-a-year business as the world's largest producer of chloroflurocarbons (CFCs). The ozone depletion issue first came to light in a plausible, although untested, theory published by Molina and Rowland in 1974. As early as 1975, then-Chairman Irving Shapiro stated that Du Pont would cease manufacturing CFCs if there was credible scientific evidence of harm to the environment.

During subsequent years, evidence of ozone depletion mounted, especially with the discovery of a growing ozone "hole" in the atmosphere over Antarctica, and scientists became increasingly convinced that CFCs were a serious threat to the ozone layer. The corresponding increase in ultraviolet radiation was expected to cause increased skin cancer and other health problems and to have other harmful effects as well—such as a warming of the earth's atmosphere, already possibly occurring as a result of the so-called greenhouse effect.

In 1986 Du Pont led industry support of international negotiations that resulted in the Montreal Protocol of 1987, which would reduce CFC production by 20 percent in 1992 and another 30 percent by 1997. The ink was barely dry on these accords when the National Aeronautics and Space Administration's Ozone Trends Panel announced new scientific findings, and at that point Du Pont's management had seen enough. The company took a leadership role by announcing a complete phaseout of fully halogenated CFCs as soon as possible, but by no later than the turn of the century. From the time it received the NASA data, the company took only seventy-two hours to chart this completely new course.

Since then the company has been involved in international meetings to strengthen the Montreal Protocol. By the end of 1990 it had invested $240 million in developing CFC alternatives, and it received the Environmental Protection Agency's 1990 Stratospheric Protection Award. As Woolard put it in his 1989 London speech:

It was an example of where we used technical data to initiate a remedy rather than to delay action. I like to think that a response of that nature represents the basic ethic of our company, and it is that ethic—one of corporate environmentalism—that will guide our future actions.

Partnerships

Partnering is more than a buzzword for Du Pont. In addition to its Plastic Recycling Alliance venture with Waste Management, the company has been involved in a number of significant partnerships. As Archie Dunham, senior vice-president of Du Pont polymers, remarked, "In the coming years, it will become more and more important to differentiate between customers who are business partners and those who are only buyers. The former are a much more valuable category, and it is important to support and strengthen these strategic relationships."

One of the most significant partnering examples is the Council for Solid Waste Solutions, of which Du Pont was one of the seven founders. Established in 1988, the Council is the plastic industry's environmental task force, with a balanced set of technical programs, government affairs, and communications activities. In 1991 it had twenty-six members, a $21 million annual budget, and a professional staff of more than forty people. As with any task force, the group will be disbanded once its task has been achieved, particularly in the recycling arena.

The Council takes the position that an integrated waste management system begins with source reduction, or minimizing disposed waste, followed by recycling, waste-to-energy incineration, and landfilling. Recycling is critical, because the Council members believe that society will not accept alternative methods of disposal until maximum value is derived from the recycling of plastics.

The Council disseminates information through a quarterly newsletter, a special publication to recyclers, an on-line news service, and a toll-free 800 number. A media monitoring service summarizes 2,000 articles a month. The organization networks closely with counterpart organizations in Europe, Japan, Australia, and Canada on technical, governmental, and communications matters. A special program for secondary school students provides information and materials relating to plastics in the waste stream and includes a handbook on how to set up a recycling program within a school.

The following are additional examples of Du Pont partnerships:

- Du Pont and the State of Illinois have entered into a major recycling partnership to develop commercial markets for PET and HDPE containers collected through curbside and drop-off recycling programs. The plastic is recycled into transportation and maintenance products, which are submitted under competitive bid to the state's Department of Transportation. For example, one of the first commercial products developed was a collapsible highway barricade made from recycled HDPE, called "SafetyCade," manufactured by WLI Industries.

- American National Can, a major producer of glass, metal, and plastic containers, is cooperating with Du Pont in a program to encourage the recycling of multilayered plastic bottles, such as syrup and ketchup bottles, and other types of plastics not currently being recycled. The cooperative development agreement began in Chicago, because of an existing Plastic Recycling Alliance facility, and involved the joint development of plastic sorting technology through the PRA as well as market development for all grades of recycled plastic.

- The Plastic Recycling Alliance and Occidental Chemical Corp. have agreed to work together to implement an automated sorting system to separate PVC bottles from the mixed post-consumer bottles at PRA's Philadelphia plant; OxyChem recycles the PVC bottles after they are sorted.

- Du Pont's packaging business and the Council for Solid Waste Solutions are working with the Flexible Packaging Association on recycling composite plastic structures and other forms of flexible packaging.

The Illinois Department of Transportation example is an excellent case study of cooperation between state government and private industry to create market pull-through for recycled products. The highway barricade project involved a large number of players on both the private and public sides, including the state's Department of Natural Resources, an industrial design firm, Waste Management, Inc., and WLI Industries, which had the design and the production capability. It is a model of cooperation that could be repeated in other states.

Du Pont has observed that many states are adopting procurement policies that would favor the use of recyclable materials or materials made from recycled content, and this represents business opportunities for the company. In some cases, states are willing to pay a 5 or 10 percent price premium as a way of pulling through recycled materials to market.

Outreach and Communications

Du Pont senior executives are active in communicating with environmental and business press, both through group and one-on-one meetings, to encourage dialogue on municipal solid waste issues and to hold themselves open to questions.

The company also works through the Council for Solid Waste Solutions to promote public education through Council-sponsored media tours, meetings with legislative bodies, hosting of regional gatherings, luncheons with civic leaders, and other activities, such as sessions with Hollywood organizations that directly influence the way environmental issues are characterized in television and movie productions. One of the more unusual meetings was with the staff of "Dear Abby" to answer environmental questions from Abigail van Buren's readers. Du Pont and a representative of the Council were the first industry representatives to meet with them and address such questions as, "Which is better, plastic or paper grocery bags?"

In addition, Du Pont meets directly with environmentalists, and sees its relationship with them changing to one that is increasingly cooperative, rather than adversarial, in nature. This is not to say that Du Pont and environmental activists are in total accord. However, the company has come to respect the thoughtfulness and sincerity of certain environmentalists and sees many points of agreement with them.

Through these several communications efforts, Du Pont is hoping to eradicate some of the "plasticphobia" that has spread through society in general, and to ensure that the industry's contributions and achievements are objectively understood.

Education

Du Pont and other member companies of the Council for Solid Waste Solutions have developed broad municipal solid waste educational programs. These are based on "educating the educators," by providing tools and funding to teachers. The Council's program is targeted at grades 6 through 12 and involves people active in the waste management field, environmentalism, and other areas not restricted to plastics.

The idea behind these initiatives is not to design curricula for teachers but to give them the tools to design their own. It is a national effort, typically on weekends, reaching audiences of 200 to 500 at a time. The Council is considering expanding the program to the college level as well, not to establish a college curriculum in waste management but to ensure a balanced viewpoint on the college campus.

Du Pont has developed and funded its own program for kindergarten through grade 6, which was introduced at the National Science Teachers Convention in March 1991. The curriculum, developed by fifteen teachers as part of the Delaware Teachers Project, will be introduced to other school districts across the country. The Du Pont program will be integrated with the Council's, to help create a complete K–12 educational program.

In Tennessee, Du Pont provided $35,000 to fund teacher training in Williamson County and to develop a municipal solid waste management curriculum. The company has awarded additional funding to support the Volunteer Public Education Trust Fund on behalf of each year's top recycling school, which will support children at risk of not completing their education. Du Pont is also awarding computers to county schools most involved in recycling.

Investing in an Infrastructure

Du Pont believes that the plastic industry is undergoing a fundamental change. Because the recycling of plastics is rapidly becoming a reality, plant capacity, utilization, and investment will likely be affected. By 1995, over a billion pounds of virgin plastic production per year may be displaced by recycled plastics.

This is some of the thinking behind the work of the Council for Solid Waste Solutions to create a national infrastructure for recycling plastics. In Phase I of its Blueprint for Plastics Recycling program, the Council has identified more than 570 plastics handlers that collect and buy bottles and containers for recycling; it had also identified 150 reclaimers and almost 100 materials recovery facilities as of December 1990. In the process, the Council has estimated an existing national capacity for recycling 620 million pounds of plastic, and it has pinpointed more than a hundred products made from recycled plastic.

In Phase II of the program, launched in March 1991, the Council disseminated this information on state-of-the-art plastics recycling to businesses and communities across the country and established specific and quantifiable recycling goals. The purpose was to make plastic recycling a reality in much less time than the aluminum industry was able to promote recycling of aluminum—indeed, to make plastic the most recyclable material in every application where plastic is used. In other words, if a beverage container can be made from glass, aluminum, or plastic, the goal is to make the plastic container the most recycled of these three types of containers.

The former director of Du Pont's Plastics Laboratory was given a leave of absence to work for the Council, to provide the experienced technical support to conceive this blueprint for a national infrastructure.

Opportunities for Leadership

With its experience in both industrial and post-consumer recycling, Du Pont is well-positioned to help provide industry leadership in solving the municipal solid waste crisis. In addition, Du Pont has established a purchasing policy that encourages its suppliers to provide materials having recycled content, including plastics, paper, and packaging. The policy has been communicated to regional purchasing agents around the world and illustrated by way of specific examples.

Directly and through the Council for Solid Waste Solutions, Du Pont also works to influence federal and state procurement policy to encourage recycling. Since the federal government is the world's largest consumer, it is a logical candidate to help create market pull through for recycling, to help achieve the EPA's goal of recycling 25 percent of the nation's waste. Du Pont has also worked with the National Governors' Association, the Council of Mayors, and the Coalition of Northeastern Governors, both through the Council and directly, to develop voluntary action plans to reduce municipal solid waste.

The Business Case for Environmentalism

Because Du Pont is one of the largest and most diversified producers of plastics in the world, it is well-positioned to help provide meaningful solutions to the municipal solid waste issue. In so doing, it anticipates new ventures and business opportunities where others might only see a problem or controversy.

Currently none of Du Pont's ventures in plastic recycling meet the company's traditional financial criteria for new projects. However, it sees this activity as a very important positioning strategy, as significant changes are taking place in the industry. Interestingly, the company had entered the post-consumer recycling business at a much earlier stage, in the early 1980s, anticipating that mandatory recycling laws would help create millions of pounds of PET feedstock for polyester uses. Du Pont sold the business because collection of bottles for use as a feedstock was below expectations. Today, the rapid growth of mandatory recycling laws and curbside recycling means post-consumer plastics are being collected on a scale that can make sense to major companies as well as to entrepreneurs.

Du Pont is also investing in new processes and technologies to complement its recycling efforts. It has patented a depolymerization process for converting PET into its primary ingredients for reuse, a process that should meet food contact criteria. The process will also handle materials that do not currently fit into existing recycling streams.

Most of the company's environmental moves cannot be economically justified as short-term return-on-investment measures, nor were they motivated by punitive legislation. For example, there are no guarantees that the depolymerization process will ever make economic sense. However, Du Pont believes that the plastics industry, like the aluminum industry, must invest in recycling to defend and preserve the long-term growth of its market. Du Pont has a number of chips on the table, from alliances to new scientific processes. In the long run, the company believes these efforts will make economic sense for the company.

The Challenges

While Du Pont is not considered a consumer goods company, consumer perceptions and misconceptions about plastics will ultimately affect the company's bottom line. As a result, Du Pont has recognized as a major challenge the need to make plastics recycling a reality.

Another challenge for Du Pont is to communicate the facts as broadly as possible, not only to opinion leaders but all the way through to the general public. This goal requires working with legislators, environmental activists, educators, opinion leaders, and consumers. Most importantly, Du Pont believes that industry has to back up its claims with performance. To this end, Du Pont is leading an industry-wide commitment and investment in a recycling infrastructure, to demonstrate that plastics can be recycled into meaningful products on a broad scale.

9

Choices for Consumers: The Procter & Gamble Approach

Few companies demonstrate a clearer commitment to environmental solutions than Procter & Gamble (P&G). There can be no doubt that the environmental movement is here to stay when a major consumer goods company of P&G's size—topping $24 billion in revenues in 1990—applies the philosophy to such established name brands as Tide, Downy, Folgers, and Spic and Span.

For P&G, identifying and understanding consumer needs is mission-critical. The company views environmental quality as a new consumer need, just as important as the cleaning power of its detergents or the taste of its toothpastes. This need for environmental quality is treated as a key business issue that is integral to P&G's global business.

P&G's Environmental Quality Policy states it this way:

Procter & Gamble is committed to providing products of superior quality and value that best fill the needs of the world's consumers. As a part of this, P&G continually strives to improve the environmental quality of its products, packaging and operations around the world.

How did this policy come about? How is it communicated throughout the company and instilled in the corporate culture? And how is policy translated into products, particularly with respect to packaging? This story begins not in Cincinnati, where the 150-year-old company is based, but in Germany.

History of Environmental Policy

Procter & Gamble had long been involved in various aspects of environmentalism. However, the company first identified environmental quality as a broad consumer need during the mid-1980s in Europe.

Particularly in Germany, the Green Party had begun to enjoy growing support that was showing up not only in legislation and regulation but also in consumer awareness. Public concern over the municipal solid waste crisis was especially acute in Germany; because of its landfill limitations, that country was incinerating three and a half times more of its municipal solid waste than the United States. In addition, most German homes were limited to two small trash cans of household waste per week: one can for mixed trash and one for recyclables. As a result, German consumers began to use their buying power to express their demands for solutions to the municipal solid waste problem.

P&G attributes its success to listening and responding to consumer needs. To answer the German consumer's concern over municipal waste, the company introduced a concentrated fabric softener in a refill pouch. Lenor, the equivalent of Downy fabric softener in the United States, was first market-tested in this form in Germany in 1986. The product meets two source reduction criteria: concentration and refillable packaging.

The pouch, made from PET with a polyethylene laminate sealant, allows the consumer to reuse the Lenor bottle. The consumer cuts off the corner of the pouch, pours the concentrate into the bottle, adds water and shakes it up to produce four liters of Lenor. The tiny pouch, rather than the plastic jug, is the only waste.

The Lenor concentrated refill pouch has now been on the market for several years and has been growing steadily, despite the fact that it is arguably less convenient than the premixed version. In fact, in P&G's own premarket tests, this packaging concept failed in every category except one: intent to purchase. P&G has found that when consumers are personally and directly touched by the environmental issue, they will voluntarily change their purchase and disposal habits.

As anticipated, the product was well-received by the German consumer, and two major German competitors paid P&G the compliment of copying the product with versions of their own. P&G's refill package also became an industry standard throughout Europe, and many other companies have created their own versions.

In 1987, it became clear that the same consumer needs were evident in Canada and the United States, and P&G anticipated that municipal solid waste management would become a potential regulatory issue as well. The company therefore introduced the refill concept to North America. Four pouch refills—Tide, Downy, Ivory Dishwashing Liquid, and Mr. Clean— were first introduced in the Canadian market; the Downy refill concentrate was first shipped to U.S. East Coast markets in the fall of 1989. The Downy product was sold in a collapsible paperboard box rather than a plastic pouch because of strong consumer preference; however, the same benefits of product concentration and refillable packaging apply.

The company has since pioneered a number of environmental concepts in the United States, including:

• Concentrated detergent powders, which require less packaging.

• Combination products, such as detergent with bleach or with fabric softener—another form of source reduction in packaging.

• Recyclable plastic packaging, as well as packaging with significant recycled content—including 25 percent recycled content in plastic film packaging for disposable diapers.

• Composting of food waste, soiled paper products, and yard waste.

To understand how these innovations came about, we should first understand how P&G is structured, and how the organization empowers its people to implement policy.

Organization and Policymaking

Procter & Gamble is a United States-based company with a rapidly growing international business. In 1980, the company had operations in twenty-two countries; by 1990, that number had jumped to forty-six. P&G is especially active throughout Europe, including Eastern Bloc countries, as well as in the Pacific Rim, particularly in Japan.

One of the company's strengths is managing change, whether it be organizational change, new products, or operations in new countries. Starting in the 1980s, P&G evolved from a highly centralized organization to one that is organized by business units. The U.S. business units are based on thirty-four product categories, such as shortening and oils, hard-surface cleaners, analgesics, and so forth. These core profit centers, the product category business units have pushed decision-making further down the organization.

In the United States, the initial environmental focus began at the middle-management level. Several middle managers identified environmental quality as a major corporate issue with far-reaching impacts for the company.

Environmentalism was a very easy "sell" to P&G top management, for several reasons. One was the company's European experience, especially in Germany. The second was the experience of then-CEO John Smale and a number of his staff, who had studied municipal infrastructure needs with the city of Cincinnati. The middle management team met with a receptive audience in Smale, who was already well-versed on the costs and implications of maintaining a municipal infrastructure.

A task force of middle management was formed to develop a solid waste policy statement; that statement, which was the genesis of the policy described at the beginning of this case study, was reviewed by top management, and it was accepted. This outcome led Smale to develop a new function called Environmental Quality.

The company's efforts to develop environmentally sound packaging began in September 1988 and became the specific responsibility of corporate packaging development. Organizationally, the challenge was to make the environment part of the menu of priorities among P&G package designers, along with containment, safety, costs, and other traditional factors. Another goal was public education—in particular, dispelling popular myths about packaging materials.

In addition, the task force quickly identified an urgent need to educate P&G employees throughout the organization, as much as it needed to educate its consumers. Some employees were still of the opinion that paper packaging was best because it was biodegradable, and such misconceptions needed to be corrected. This was accomplished through numerous presentations, articles in in-house publications, and reports on degradability.

In short, corporate empowerment for environmental policy began with an initial concept from middle management, who had observed the company's German experience and anticipated similar change in the United States. These middle managers agreed upon a policy statement that was delivered to top management. An informed and receptive CEO approved their position, which was then refined and distributed as corporate environmental policy. A central task force was given responsibility for implementing the policy and a specific Environmental Quality function was established and staffed. The task force has since evolved into a coordination effort, with direct responsibility for environmental solutions transferred to and embedded in the business units.

In this way, P&G applied a fairly traditional approach for policy formation, from committee recommendation to business practice. The committee or task force approach begets the expertise and the depth that is needed, upon which a leadership role can be based. Most importantly, P&G determined that environmental quality cannot be a special project; it had to be part of the way P&G does business.

Plastic Packaging Initiatives

P&G recognized that plastic was eventually going to be a serious issue in the public's eyes. Plastic was a problem not from a technical perspective but from a public perception standpoint. The public mistakenly believed plastic was unrecyclable and dangerous to incinerate. The company had to educate the consumer and to develop sound environmental policies with respect to plastic *before* these negative perceptions escalated into full-blown consumer rejection.

In the light of this anticipated public resistance to plastic packaging, P&G wanted to take some proactive and preemptive measures. Philosophically, P&G wished to be a world leader in this area and assume a very high profile. This is very uncharacteristic for P&G, a traditionally low-key and conservative company. P&G had always closely guarded its corporate strategies and practices, and it had certainly never publicized planned initiatives in advance. With environmental products and packaging, the approach would be markedly different.

P&G had learned some lessons from the phosphate controversy of the 1960s. From its scientific studies, P&G had concluded that phosphates in detergent had little to do with municipal water pollution. The company had believed that truth, based on the technical facts, would prevail, but in this case, regulatory and special interest perceptions proved more powerful than fact. As a result, P&G ended up providing a phosphate-free detergent in those areas of the country where phosphate content was regulated.

The lesson was clear: Public perceptions about plastic packaging needed to change. And so, in November 1988, the company unveiled its Spic and Span Pine Cleaner in a clear bottle made from 100 percent recycled PET. In a major departure from normal practices, P&G had announced the packaging concept long before it went into regular production. The company had made some preliminary production runs to verify the technology. At that point, no other company had delivered a major consumer product in a completely recycled plastic container. It was a major triumph for P&G.

P&G's efforts were rewarded with what has been one of the company's most successful new packages. The Spic and Span Pine bottle, made by PlastiPak Packaging, Inc., has won a number of packaging awards, including the IoPP Packaging Innovation of the Year, the Du Pont Environmental Award for Packaging, and awards from Renew America and Keep America Beautiful.

Now generally available and actively marketed, the Spic and Span Pine Cleaner in the 100 percent recycled PET bottle has been the beneficiary of a virtual avalanche of consumer testimonials. Under any circumstances, it is unusual for a consumer to call or write to a company to praise a floor cleaner. Instead of the usual two or three letters a year, P&G was now receiving hundreds of consumer letters and calls praising the new bottle and thanking the company for its environmental responsibility.

In the wake of this positive response, the company turned its attention to high-density polyethylene, or HDPE. Most of its plastic packaging was HDPE (more than 100 million pounds of HDPE a year go into P&G containers), so P&G called in three of its plastic bottle suppliers to look for new approaches to the material. Each company provided a characteristically different response.

In May 1989, P&G held a joint press conference with all three bottle suppliers, at which each supplier demonstrated the P&G containers it produced with recycled plastic. The Tide, Downy, and Cheer containers had 20–30 percent recycled content. This announcement marked the beginning of large-scale plastic recycling, and it made headlines.

Not all of the technical issues had been resolved at the time of the announcement. One of the issues in using recycled HDPE is color fidelity. Dyes are expensive, and it is difficult to achieve "Cheer blue" or "Tide orange" with recycled material. Tints vary in recycled HDPE even if it is unpigmented, ranging from white to gray.

However, as the company predicted, the biggest issue turned out to be odor. For example, the HDPE milk bottles that are a part of the recycled plastic supply may have a sour milk odor. Product odor is, of course, critical to product design in this field.

These issues were resolved with a three-layer construction for the HDPE containers. The innermost layer, which comes in contact with the product, is virgin plastic, so as to protect the integrity of the product. The outermost layer is also virgin plastic and contains the bright color dyes. Only thin layers of virgin material are needed for these purposes. Finally, the middle layer is recycled HDPE, made from both recycled plastic and regrind from the bottle-making process.

The company now has several major products in HDPE bottles using recycled materials, including Tide, Downy, Cheer, Era, Bold, and Dash.

Another P&G packaging innovation is the first coffee jar made from plastic that can be recycled. In 1990, the company began test-marketing its Folgers instant coffee in a clear PET jar, in addition to offering the steel coffee can and the vacuum-packed brick bag. The PET coffee jar has no blaring "recyclable" claims on the label; instead, an informational tag attached to the neck of the jar explains that the jar can be recycled along with PET soft drink bottles. In addition, P&G researched in advance all the recyclers across the country that would accept the PET jar, and makes its list available to consumers who call the company's toll-free hotline. P&G believes that customers will appreciate the informative, "non-salesy" approach it has taken.

The company makes no attempt to rank the environmental compatibility of the PET jar against that of the steel can or brick bag. Each package has its merits. If the consumer is committed to recycling and has access to PET recycling facilities, the PET jar is a good choice. The same is true of the steel coffee can. However, if the consumer has no plans or opportunities to recycle, the vacuum bag—currently a nonrecyclable composite structure made from aluminum foil and polyethylene sealant—is a good example of source reduction and will take up significantly less space in landfills. P&G thus assumes the role of making environmental choices available, but the final decision on the post-consumer fate of the package is a function of the available infrastructure.

While nonrecyclable packaging can yield significant source reduction, P&G sees this alternative as only an interim solution. In the long term, the company expects all of its products and packaging to be compatible with recycling or composting.

Building an Infrastructure

When P&G first made a commitment to recycled content plastic packaging, very little recycled material was available. The first priority was to establish a plastic recycling infrastructure. As a first step, P&G joined the Council for Solid Waste Solutions as the first nonpetrochemical industry member. High on the company's agenda was the goal of creating a recycling infrastructure for HDPE—the material used most in P&G plastic packaging.

As P&G saw it, the best way for the company to create that infrastructure was to be the market for recycled plastics. As the Corporate Packaging Group put it, the first three priorities in creating any recycling system are markets, markets, and markets. And if P&G was not buying the recycled materials, it was in no position to suggest that anyone else do so.

P&G dispatched staff to plastic recyclers to inspect their recycled material, to determine its quality and availability, and to assess the quantities that

P&G would need. P&G does not buy plastic scrap itself, but it creates demand and provides performance specifications to the plastic bottle suppliers. In the past, P&G would not have concerned itself with the bottle-makers' source of supply. However, because of P&G's high profile in publicizing plastics recycling, the company became much more involved in the supply of recycled resin than it normally would have been.

Partnerships With Consumers

P&G believes that its greatest impediment is uninformed public perceptions. For example, the company carefully researched public opinion on plastics, and it found that consumers were uncomfortable about the material. On the one hand, they indicated to P&G that they liked many aspects of plastic, such as the fact that it is light-weight, unbreakable, inexpensive, and colorful. However, they also "felt bad" about plastics. They were troubled by problems they perceived in disposing safely of plastics. They did not like the fact that plastics are made from petroleum, a nonrenewable resource. For the most part, their feelings were not founded on technical fact.

Degradability is another poorly understood issue. Few consumers know that virtually nothing degrades in a modern landfill. While informed consumers are gradually learning that degradability is no silver bullet, the myths about it are persistent. As recently as 1991, legislation was still being introduced calling for a ban on nonbiodegradable packaging materials.

P&G's approach to working with consumers is to be responsive to customer calls and information requests, and not to irresponsibly exploit environmentalism as a competitive issue. People want to know how to be environmentally responsible, and the company tries to be as informative as possible and to avoid exaggerated or misleading claims. P&G concluded that the best advertising claim tells the truth about the product in a specific way. The label on the Spic and Span Pine PET bottle simply says that the container is made from 100 percent recycled plastic. On the back of the bottle, the label displays the SPI "chasing arrows" symbol and devotes a small paragraph to describing the environmental qualities of the package.

Here too, the company favors a factual, informational approach, as opposed to "sell" copy, on its labels.

P&G nationally advertises its environmental packaging, both the Spic and Span Pine bottle and the Downy refill. The company also has consumer 800 numbers in place to answer customers' questions about environmental packaging and products.

P&G has supported the development of television public service advertisements that deal with environmental issues in general, especially plastics recycling. Through involvement with Closing the Loop, a public/private partnership in the Washington, D.C., area, public awareness of the importance of recycling and participation in local recycling programs has improved. Some of the participants in this project include local public works and waste management departments, a grocery store chain, environmentalists, Virginia Governor William Schaeffer, District of Columbia Mayor Sharon Pratt Dixon, and local TV stations. Closing the Loop is also working with local schools and community groups to provide educational materials on recycling.

The consumer's confidence in P&G is critical to the company. P&G sponsored a 1989 survey that dealt with national consumer attitudes about the environment. One of the questions P&G asked was, "If you want to know the truth about solid waste and the environment, where would you go?" The responses were fairly evenly divided between the media, universities, and government. Not one, out of several hundred consumers, mentioned turning to industry.

In P&G's experience, consumers are rapidly changing behavior patterns in response to environmental concerns—in fact, faster on this issue than on any other. The breadth of the change remains to be seen. However, P&G believes environmentalism is more than a niche behavior. Consumers are buying products on a broad basis of environmental criteria.

P&G also is trying to change public perceptions by working with educators. A number of P&G professionals consult with school counselors, teachers' organizations, and civic groups that have access to schoolchildren, and devote personal time to delivering speeches to these groups. The company believes children are one of the sources of environmental information for the home and can help to encourage household sorting and recycling. Another example of P&G's commitment to educate young people about the environment is Planet Patrol, an environmental teaching unit about solid waste for upper elementary grades. The program focuses on the EPA's hierarchy of integrated waste management solutions and is made available to teachers through professional meetings, P&G catalogues and direct requests.

Working With Retailers

The company also works with the supermarkets and other retailers of its products. Some request that P&G provide a list of its environmental "good guys," but P&G prefers not to operate that way. Instead, it works with trade associations like the Grocery Manufacturers Association (GMA) and the Food Manufacturing Institute (FMI). The company wants to help the retail stores develop an ability to respond to public concerns on a broad basis, rather than merely be a conduit for consumers' demands and manufacturers' claims. Some of the company's products inevitably show up in supermarkets' "green campaigns"; some of these campaigns meet with P&G's approval, while others are deemed rather specious.

P&G's concern is that product acceptance programs do not contribute to long-term solutions. In order to achieve long-term solutions, all of the following three things must happen:

1. Consumers must change their buying habits.
2. Consumers must change behavior patterns to participate in composting and recycling programs.
3. Consumers must understand the issues enough to support the needed infrastructures.

Retailers' product acceptance programs may help change buying habits, but they do not contribute to consumers' understanding of the issues so as to change their behavior patterns or to influence their vote on municipal solid waste issues; in the end, the programs may actually retard progress on long-term solutions. Instead, P&G believes the solution is clear labeling guidelines, standards, and definitions by which the consumer is educated to make informed choices.

A growing number of retailers in the United States, are taking a proactive stance in driving the environmental movement. P&G is establishing partnerships with these large chains, as well as with distributors that service independent retailers.

P&G and its retail customers have joined with Keep America Beautiful (KAB) and its local affiliates in a program called "Lets Not Waste the 90's." This program is also designed to help educate consumers on solid waste management alternatives. Retailers work with their local KAB affiliate to create educational programs designed specifically for their community. The programs may range from the organization of a recycling Awareness and Pledge Day to an essay contest or sponsorship of a school play. In conjunction with this program, P&G ran a TV commercial followed by a coupon insert in the following Sunday's newspapers. The importance of this program is that it gives tips to consumers about how they can make a difference through their own actions, and it encourages retailers to get involved in their communities at the grassroots level.

Influencing Suppliers

P&G believes that buyers like itself, as well as retailers, warehousers, and private labelers, can drive the trend toward recycling by specifying mass volumes of recycled packaging from suppliers. Its work with plastic bottle suppliers, as described above, is instructive. The company directly influenced the recycling infrastructure and made the suppliers a visible part of its public awareness campaign.

By collaborating closely with its suppliers, P&G proved that plastic recycling can be achieved. And what is more, recycling can work with major P&G brands. These are not isolated experiments on low-profile brands, but large-volume, flagship products. Tide is the top-selling brand of the detergent industry, and Downy is the top brand of fabric softener. Plastic recycling was no longer a laboratory curiosity.

In addition, P&G's powdered detergents are sold in recycled paperboard boxes. More than 80 percent of the 1.5 billion paper cartons P&G produces each year are made from recycled fiber. P&G indicates it could use even more recycled material, and is limited only by availability. The company continues to field calls from potential purveyors of used HDPE bottles and refers them to its plastic bottle suppliers.

Working With Legislators

P&G has long had a system for monitoring legislation. And because of P&G's high profile, it is often invited to participate with joint public/private groups. Since 1988, P&G has been active in the Coalition of Northeast Governors (CONEG), Source Reduction Council, where it has worked closely with representatives from the 9 northeastern states, private industry, and leading environmental groups.

P&G believes that educating legislators is as urgent a need as educating consumers, retailers, industry, and activists. The company is concerned that source reduction is becoming eclipsed by recycling in the regulators' eyes. For example, the company's coffee brick bag creates 85 percent less trash than tin cans, but it may face bans and punitive taxes because it is not considered recyclable. The company believes that large source reductions are preferable to low recycling rates in the short term. However, P&G expects that all its products and packaging will ultimately be compatible with recycling or municipal solid waste composting.

P&G's work with legislators exemplifies its conviction that industry is best positioned to take a leadership role in environmentally sound packaging. Legislators need to be informed by industry, but this is an uncomfortable and uncharacteristic role for most corporations. P&G has been aware of this, and as a result of taking a leadership role, has been able to have significant influence in groups such as CONEG. For example, CONEG's heavy metals bill is derived directly from P&G policy.

Some companies have accused P&G of helping legislators regulate industry. The company is involved in a CONEG caucus of industry members, and it has learned that some businesses, who are not particularly involved in the issue, cannot comprehend why P&G "consorts with the enemy." P&G finds that some companies are still uneasy working in partnership with legislators, seemingly on an emotional or "gut" level.

Industry Cooperation

Through one-on-one meetings and industry forums, P&G has urged other companies to voluntarily support initiatives such as CONEG's Preferred Packaging Guidelines. The company also works with numerous trade associations, such as the FMI, the GMA, and the National Food Processors Association (NFPA). It is an active member of the Council for Solid Waste Solutions.

P&G was also one of several companies that established the Solid Waste Composting Council. The company believes municipal solid waste composting is the best disposal solution for a great deal of paper-based materials, including its paper diapers. Composting is increasingly viewed not only as an important municipal waste management technology, capable of diverting 30 to 60 percent of today's garbage—food, yard waste, and soiled paper—but also produces a useful end product that can help enrich and add life to our nation's soil. Composting is helped by the fact that there are few negative perceptions about it. It is understood as a natural, organic process. And for large portions of the country, composting is an economically attractive alternative to landfilling or incineration. While the industry is in its infancy, with fourteen facilities in operation in the United States in mid 1991, ten new facilities planned for a total of 24 by the end of 1991 would more than double 1990 production rates. At the same, time, more than 150 municipal composting projects are in various stages of planning.

By working with industry associations—and developing new ones—the company "gets the message out" to industry in a variety of ways.

Working With Activists

P&G has had mixed experiences working with environmental activists. Some *cannot* work closely with industry because of their organizational charters. Others decline to do so for political reasons. Some environmentalists also face a kind of peer pressure that discourages them from "consorting with the enemy." Membership and funding in environmental groups may depend on their playing the role of outsiders, and this flies in the face of collaboration with industry. And some environmentalists feel their objectivity may be questioned if they are perceived as endorsing the policies or products of a corporation.

The company respects the technical expertise and personal integrity of a number of environmentalists, and its work with environmental groups has been beneficial, overall.

Market Positioning

P&G expects that if it meets the consumer need for environmental quality better than anyone else, it will achieve a competitive advantage. At the time of this book's publication, however, it was too soon to provide hard evidence of a change in market share. The company had just begun to distribute nationally its Downy refill pouch and its Spic and Span Pine in the recycled bottle.

Instead, the company can point to its testimonials which, as noted earlier, have been overwhelming in response to its packaging initiatives. Customers have indicated that they will switch to P&G's Spic and Span Pine brand because of the recycled bottle, although it remains to be seen if purchase decisions match intentions.

It may not even be possible to validate the competitive advantage of having environmental packaging because of the large number of factors that affect sales in a highly competitive consumer product business. Nevertheless, P&G is pursuing the basic notion that those who best satisfy the needs of the public will prevail.

It is interesting to note that P&G, like other companies leading this movement, is implementing environmental strategies without relying on strict return-on-investment criteria. P&G has placed environmental quality on every packaging designer's plate and is making sure the company is doing the "right thing."

P&G believes it will cost no more, and in some cases it will cost less, to implement environmentally sound packaging. At the same time, the company makes it clear that it will not compromise the integrity of the product and will continue to provide packaging that keeps its products fresh and safe, prevents damage, and provides tamper resistance. The company will stay on course with its environmental position, and it is convinced it will build market share in the process.

P&G's goal in working with retailers, suppliers, legislators, and activists is ultimately to make responsible choices available to consumers, and to help them make sense of the trade-offs. Because of its ability to influence suppliers through its purchasing power, and to influence consumer behavior through marketing and educational activities, P&G is well-positioned to foster environmental responsibility all along the supply-and-demand chain and to influence the direction of packaging. P&G shows that industry, through its unique experience and knowledge of the issues, can play a decisive role in creating a partnership for progress.

10

Johnson & Johnson: Worldwide Environmental Responsiveness

Johnson & Johnson (J&J), with over 83,000 employees worldwide, is the largest and most comprehensive manufacturer of health care products, serving the consumer, pharmaceutical, and medical products markets. The company is one of the world's most environmentally responsible companies; in *Fortune* magazine surveys, American business leaders ranked J&J number one in community environmental responsiveness three years in a row. How does J&J, as a highly decentralized organization with farflung units around the globe, maintain its commitment to the environment ?

Part of the answer lies in the J&J corporate culture, which adheres to its longstanding Credo of responsible corporate behavior. Examples of the Credo in action may be seen in the company's policy of openness and information sharing with the local communities where it operates. This includes donating innovative software and communications tools, as well as facility maps, to local authorities, to help handle chemical emergency response situations. (The company's applications of information technology are described below in further detail.)

Another critical factor is the decentralized organizational structure of this large multinational company, which facilitates the communication and implementation of responsible environmental policy across a wide array of geographic markets and product sectors. This chapter begins by examining J&J's corporate structure.

Corporate Organization

The company is organized on the principle of decentralized management, with an executive committee responsible for the operation of the corporation. With its corporate headquarters in New Brunswick, New Jersey, J&J is made up of approximately 175 essentially autonomous companies with over 200 facilities in 54 countries. The largest concentration of facilities is in the United States and Europe.

The entire corporation worldwide is divided into three distinct sectors along both product lines and market focus:

• The Professional Sector produces sutures, mechanical wound closure products, diagnostic products, and other related items, used principally in the professional fields by physicians, dentists, hospitals, laboratories, and clinics.

• The Pharmaceutical Sector produces prescription drugs, including therapeutic medicines and antifungal products. There are also large research and development facilities in this sector.

• The Consumer Sector produces toiletries and hygiene products, including baby care items, first-aid products, and nonprescription drugs.

Each sector has an operating committee which oversees and coordinates the activities of domestic and international companies. The operating management of each company is headed by a president, general manager, or managing director who reports directly to a company group chairman. As a decentralized corporation, policies and programs may be implemented through the corporate office, through companies, or through individual facilities.

The J&J Credo

How does such a diversified, decentralized, worldwide company find single purpose and answer the challenge of complex global environmental issues? As Dr. Peter Britton, Director, J&J Community Environmental Development, put it, "Our management philosophy is based on our Credo, a document written fifty years ago by General Robert Wood Johnson. It has served us well these many years from the Tylenol crisis to the way we manage current environmental issues."

The Credo (see Figure 10-1) sets forth J&J's responsibility to four constituencies: its customers, its employees, the local communities where it operates, and its stockholders—in that order of priority. J&J's handling of the Tylenol crisis is an example of the Credo in action. When bottles of Tylenol were found to be laced with deadly cyanide, the company acted swiftly to protect its customers by pulling the pain reliever product from store shelves nationwide. Indeed, the company's decentralized structure proved to be an advantage in this case. All of J&J's sales forces—not just McNeil Consumer Products Company, the makers of Tylenol—worked together to make this happen. The company's decisive handling of the crisis, and its subsequent moves to develop tamper-resistant packaging, have stood as a model to other companies.

From the very beginning, the Credo has provided a strong foundation for the company's commitment to the environment. According to Dr. Britton, the strength of J&J's worldwide commitment to the environment rests in the third point of the Credo: "We are responsible to the communities in which we live.... We must maintain in good order the property we are privileged to use, protecting the environment and natural resources."

Over the years, the Credo has been modified with the changing times. In 1979, the reference to "protecting the environment and natural resources" was added. Since that time, the environment stands prominently as a management commitment at each of the more than 200 J&J facilities worldwide.

Figure 10-1. The J&J Credo.

Our Credo

We believe our first responsibility is to the doctors, nurses, and patients,
to mothers and fathers and all others who use our products and services.
In meeting their needs, everything we do must be of high quality.
We must constantly strive to reduce our costs
in order to maintain reasonable prices.
Customers' orders must be serviced promptly and accurately.
Our suppliers and distributors must have an opportunity
to make a fair profit.

We are responsible to our employees,
the men and women who work with us throughout the world.
Everyone must be considered as an individual.
We must respect their dignity and recognize their merit.
They must have a sense of security in their jobs.
Compensation must be fair and adequate,
and working conditions clean, orderly, and safe.
We must be mindful of ways to help our employees fulfill
their family responsibilities.
Employees must feel free to make suggestions and complaints.
There must be equal opportunity for employment, development
and advancement for those qualified.
We must provide competent management,
and their actions must be just and ethical.

We are responsible to the communities in which we live and work
and to the world community as well.
We must be good citizens—support good works and charities
and bear our fair share of taxes.
We must encourage civic improvements and better health and education.
We must maintain in good order
the property we are privileged to use,
protecting the environment and natural resources.

Our final responsibility is to our stockholders.
Business must make a sound profit.
We must experiment with new ideas.
Research must be carried on, innovative programs developed
and mistakes paid for.
New equipment must be purchased, new facilities provided,
and new products launched.
Reserves must be created to provide for adverse times.
When we operate according to these principles,
the stockholders should realize a fair return.

Environmental Policy

Until the mid-1980s, each J&J company was responsible for adhering to the Credo in its own way with regard to the environment. There was no special corporate focus or directive for implementation. As the need for an explicit environmental policy became evident, the company saw the requirement for corporate direction. As a result, the company retained two third-party firms to perform the company's first worldwide corporate environmental audit. Subsequently, the company identified many changes to make its environmental program more quantitative.

In 1987, J&J strengthened its environmental commitment with the adoption of the J&J Worldwide Statement on the Environment (see Figure 10-2). This statement serves as the driving force for the company's current global initiatives. The following two major corporate programs have emerged: (1) Environmental & Regulatory Affairs and (2) Community Environmental Responsibility.

The Environmental & Regulatory Affairs program covers the technical aspects of the environmental initiatives taking place at J&J companies worldwide. It further addresses compliance with regulations and corporate management directives, the development and monitoring of environmental policies and the environmental quality assessments (audits) of facilities worldwide.

The Community Environmental Responsibility program addresses J&J's overall responsibility for ensuring that plants, products, and processes have minimal, and ideally neutral, environmental impact. Additionally, it is committed to enhancing communications of this philosophy to J&J employees, the local communities in which they operate, and to the world community at large. The intent is to develop partnerships toward the goal of working and living in harmony with the environment.

J&J's environmental program is managed at the highest level: the Office of the Chairman. The Office of the Chairman consists of the chief executive officer and two vice-chairmen, one of whom has been assigned specific responsibility for overseeing the environmental program and serving as chairman of Environmental Cross Sector Task Forces. At this writing, there are two task forces, one responsible for North America, and the other for European Operations. Each task force is composed of representatives from

Figure 10-2. J&J Worldwide Statement on the Environment.

POLICY

To protect the environment and natural resources related to all Johnson & Johnson operations worldwide through responsible management exercising excellence in environmental control.

It is the responsibility of every employee to conscientiously observe Johnson & Johnson's environmental policy and guidelines designed to protect air, land, and water and assure proper waste disposal.

Responsible environmental protection is one of the priority areas of accountability against which management performance is measured.

The managements of all companies, operating units, manufacturing plants, and research laboratories are expected to ensure their operations and facilities meet the requirements of the Corporate policy and guidelines, as well as all applicable laws and governmental regulations.

In so doing, they will:

- consider and address environmental impact in:
  - the design and construction of new and existing products, processes, equipment, and facilities
  - the selection and use of raw materials
  - the acquisition or lease of property
  - the discontinuation of operations and disposition of property
- minimize the generation of hazardous wastes and satisfy corporate and legal requirements in their disposal
- establish and maintain procedures for handling environmental incidents and emergencies
- train employees in proper handling of hazardous substances
- adopt company operations and practices which do not create environmental hazards, where laws are non-existent or less stringent than corporate guidelines/standards
- conduct audits and correct any environmental control deficiencies identified

Corporate staff will assist operating companies in the implementation of this policy as follows:

- develop appropriate guidelines and procedures to assist in the implementation of corporate policy and the law
- establish and maintain liaison with appropriate governmental agencies
- provide training and consultation to operating company personnel
- audit facilities for compliance
- report annually on environmental issues, programs, and compliance

the three operating sectors: Law Department; Public Affairs; and Health, and Safety and Environmental Affairs. In a recent development, the J&J board of directors appointed a Public Policy Advisory Committee responsible for reviewing the corporation's policies, programs, and practices on public issues regarding the environment, as well as health and safety. The Committee has also been given the authority to make recommendations to the board concerning environmental issues of importance to the corporation.

While environmental initiatives are the responsibility of each autonomous company within the corporation, the two programs, Environmental & Regulatory Affairs and Community Environmental Responsibility, provide the J&J family of companies with corporate guidance as required.

Environmental & Regulatory Affairs Program

Initiated in 1987, the Environmental & Regulatory Affairs Program is supported by a worldwide third-party environmental quality assessment (audit). This concept includes tools and guidelines which are incorporated into the *J&J Worldwide Environmental Practices Manual.* The Manual contains the basic requirements that satisfy the intent of the company's Credo and the Statement on the Environment and that were developed to standardize the environmental programs at all its facilities worldwide.

The *Environmental Practices Manual* is a reference guide used in conjunction with the J&J Worldwide Environmental Quality Assurance Program. Here, each facility is evaluated to ensure that it is in compliance with local, regional, and national environmental laws and regulations. This assessment program provides J&J facilities around the world with a third-party independent review and evaluation of all environmental programs to ensure that they are in compliance with governmental regulations, J&J policies and guidelines, and good environmental practices. The audit program is completely computerized. The auditor uses a laptop computer to record his or her findings and then prints out a formal management action plan that is discussed at the closing conference. A major effort during the

Environmental Audit Program is to provide technical services to the facility management as well as raise awareness of regulatory issues.

Policies that support the program are outlined in Figure 10-3 and include the following:

• The J&J Worldwide Storage Tank Policy includes guidelines that establish standards for proper management of storage tanks. In support of this policy, J&J, in 1988, initiated a five-year program to upgrade all its storage tanks to meet the J&J standards. The objective is to move all underground and on-ground tanks to above ground so that any leakage is easily observed. This will be accomplished by the end of 1992.

• The Biological Waste Policy requires all biological waste to be rendered noninfectious and unrecognizable prior to disposal. Biological waste as defined in this policy includes all needles and syringes, both used and unused.

• The Carcinogen Policy requires all J&J companies that use such materials in any of its manufacturing or laboratory facilities to find suitable substitutes. If noncarcinogenic substitutes cannot be found, the policy requires the installation and operation of release control devices during 100 percent of the process operation time.

Figure 10-3. J&J environmental/regulatory affairs policies.

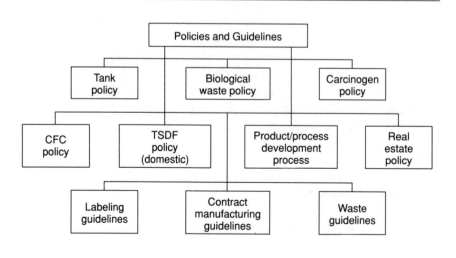

• The Chlorofluorocarbon Policy, issued in July 1989, required the removal of chlorofluorocarbons or CFCs from J&J products and all manufacturing processes by January 1, 1990. The company achieved that goal except for sterilization operations and those products and processes requiring governmental regulatory approval/oversight. Even in these processes, the company is committed to CFC elimination by January 1, 1994.

• The Domestic Hazardous Waste Treatment, Storage, and Disposal Facilities Policy applies to J&J facilities in the United States. It requires all domestic affiliates to use only those off-site treatment, storage, and disposal facilities on a J&J approved list. This policy was originally adopted because of the high liability associated with improper disposal of hazardous waste in the U.S. However, this principle is being reflected in new environmental legislation being written throughout the world.

• The Product/Process Development Policy was approved in 1990 and requires that an environmental impact assessment be performed on all research and development projects. Raw materials, processes, packaging, and final products must be taken into account in order to ensure minimal environmental impact and, ideally, environmental neutrality.

• J&J's Real Estate Acquisitions and Divestitures Policy, its most recent policy initiative, requires each J&J company to arrange for a prepurchase or a divestiture environmental audit on all real estate property to ensure that environmental liabilities are identified.

In addition, the company is developing guidelines on product labeling, contract manufacturing and the management of waste.

The Environmental and Regulatory Affairs program is very comprehensive and is committed to issuing policies and directives, assessing manufacturing sites for environmental and regulatory compliance, and most importantly, providing guidance to J&J employees throughout the world on environmental issues and concerns.

Community Environmental Responsibility Program

The overall goal of the Community Environmental Responsibility Program is to ensure that each facility worldwide operates at the same high level of environmental preparedness. The focus of this approach is to foster employee and community trust in the environmentally responsible nature of J&J products and processes, and to communicate environmental information in an effective manner to employees and to the members of the community in which the company operates.

The Community Environmental Responsibility Program has four major phases that can be further divided into steps (see Figure 10-4). Phase I relates to the training of facility managers in crisis management and is directed toward awareness, preparedness, and communication. In this phase, key facility management personnel worldwide attend a three-day training program in order to increase their awareness of environmental issues, as well as develop a high level of emergency preparedness. Additionally, it is directed toward increasing facility participation and interaction with the local communities.

The program consists of hands-on communications training through participatory workshops where tools are introduced that will prove to be beneficial at the company's facilities. The training consists of the following elements:

- An overview of the J&J Corporate commitment to the program
- Background on worldwide environmental issues and worldwide regulations
- Training on how to communicate chemical information successfully to the media and public
- Planning for a crisis by stressing the importance of being prepared for incidents involving potentially hazardous chemicals

In addition to the training, each manufacturing location is provided with a *Chemical Communications Handbook* and a *Facility Incident Manual*. The *Chemical Communications Handbook* is designed to help facilities address environmental and safety inquiries from the public. The handbook also includes definitions and characteristics of hazardous or toxic chemicals, as well as safety measures and emergency response procedures.

The purpose of the *Incident Manual* is to assist facility management in preventing and, if necessary, responding to a hazardous chemical incident.

Figure 10-4. J&J's community environmental responsibility program.

Training/Preparedness

Step 1	Step 2	Step 3	Step 4	Step 5
Data Collection and Analysis	Training/ Compliance/ Communication	Facility Emergency Response Plan	*Facility Incident Manual*	Annual On-Site Incident Simulation and Training

Facility Action/Community Communications

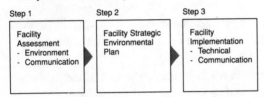

Step 1	Step 2	Step 3
Facility Assessment - Environment - Communication	Facility Strategic Environmental Plan	Facility Implementation - Technical - Communication

The Marketplace

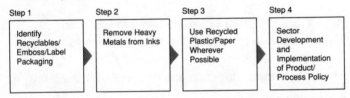

Step 1	Step 2	Step 3	Step 4
Identify Recyclables/ Emboss/Label Packaging	Remove Heavy Metals from Inks	Use Recycled Plastic/Paper Wherever Possible	Sector Development and Implementation of Product/ Process Policy

Public Affairs Activities

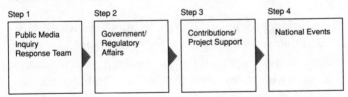

Step 1	Step 2	Step 3	Step 4
Public Media Inquiry Response Team	Government/ Regulatory Affairs	Contributions/ Project Support	National Events

This includes minimizing the potential impact of a chemical incident on the facility, its employees, and the surrounding community.

At the training session, there is a demonstration of J&J's newest tool, the J&J Worldwide PC-Based Regulatory, Environmental, Chemical Information System, or "PRECIS." This computerized system integrates chemical and regulatory data on hazardous materials stored on site with emergency planning and response information (see Figure 10-5). With the ever increasing volume of information required under present and proposed environmental legislation, the system has become virtually a requirement in order to manage chemical data, Material Safety Data Sheets (MSDS), and emergency planning and response information. The system is also designed to be used when conducting drill simulations with local emergency responders, as well as managing an actual crisis situation.

While J&J developed the PRECIS system in-house, an innovative emergency management component called EIS/C (Emergency Information System/Chemicals) was developed for J&J by an outside contractor. EIS/C combines three elements: data, communications technology, and an electronic mapping capability that allows the company to map the locations of regulated chemicals. J&J has donated the EIS/C software, as well as personal computers or the funds to purchase them, to local emergency management authorities including fire fighters and police. In this way, both J&J and the community can be well prepared for potential chemical emergencies.

Figure 10-5. J&J Worldwide PC-Based Regulatory, Environmental, Chemical Information System.

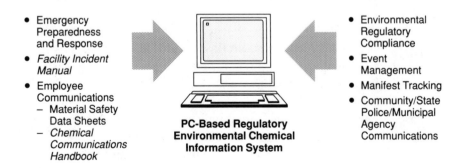

- Emergency Preparedness and Response
- *Facility Incident Manual*
- Employee Communications
  - Material Safety Data Sheets
  - *Chemical Communications Handbook*

PC-Based Regulatory Environmental Chemical Information System

- Environmental Regulatory Compliance
- Event Management
- Manifest Tracking
- Community/State Police/Municipal Agency Communications

What typically happens in the event of a hazardous chemical emergency is that the local authorities have no way of pinpointing where the accident has occurred, what other chemicals are stored on-site, and other vital information needed to address the accident. For this reason, the mapping capability of J&J's emergency response system is especially useful. Besides depicting the J&J facilities and the location of chemicals, the maps depict the surrounding community, showing the location of schools, transportation systems, and other important sites. The system also collects real-time meteorological data from local weather stations to plot the dispersion plumes of airborne chemicals.

As of 1991, EIS/C was pilot-tested at eight sites in New Jersey as well as several sites in Portugal, Belgium and the UK, and would eventually be expanded to additional sites worldwide. This information sharing is part of J&J's policy of community involvement.

The incident simulation or drill is the last step in Phase I of the J&J program. This can be conducted on three levels of complexity with Tier I involving the facility and Corporate Headquarters, Tier II adding first responders, and Tier III involving the community and media as well. The company may be the first to have conducted international chemical emergency drills; in May 1991 a J&J facility in Belgium transmitted live data on a simulated chemical spill to corporate offices in New Jersey, where the mock disaster was monitored as it happened via modem hookup.

The combination of these tools, in addition to the training, provides management with the resources necessary to help make informed and responsible decisions. J&J feels very strongly that it must make the concerted effort to work in harmony with the environment and the communities in which it operates. The company sees this commitment as good practice and good business.

By the end of 1990, key management personnel at facilities located in the United States, Canada, the Caribbean Basin, Europe, Australia, New Zealand, and India had undergone Phase I training. Phase I training was then scheduled for personnel at facilities in the Far East, Latin America, Central America, Mexico, and Africa.

Phase II, Facility/Community Communications, has the objective of establishing environmental leadership in employee/community communications and technical process technology. This is done through the use of state-of-the-art process technology to control and minimize the environmental impact of products and processes. A key component of Phase II is the development of a facility-specific environmental strategic plan whereby steps are taken to minimize the impact of various processes on the environment through continuous improvement. This phase also stresses enhanced employee/community communications, as well as assistance in community environmental programs.

Phase III, The Marketplace, has the objective of making J&J the producer of products that are environmentally neutral throughout their life cycles. This is recognized as a long-term goal that can only be achieved through the quality improvement process of "continuous improvement." The major steps in this phase are the removal of heavy metal printing inks from packaging, and the use of recycled materials, whenever possible, in both products and packaging. J&J products will provide information which can be in the form of a label or a product insert, to assist customers and users in making an informed, environmentally sound, decision in the marketplace.

A major thrust of Phase III is the implementation of the Product/Process Development Policy. A key component of this policy is the use of environmental impact assessments or life cycle analyses on new products and processes, as well as on major existing products, in order to help minimize their potential environmental impact (see Figure 10-6). In implementing the Product/Process Development Policy, the total impact of a product from

Figure 10-6. J&J Product/Process Development Policy

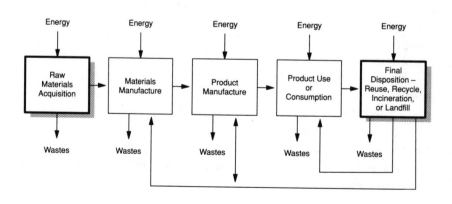

"cradle to grave" is evaluated through a cooperative effort of marketing, operations, and research personnel. It is through this policy that J&J is committed to achieving environmental neutrality.

Phase IV, the final phase of the Community Environmental Responsibility Program, is Public Affairs Activity. The goal is to take a visible role in shaping environmental initiatives in the civic arena. A number of programs are already in place relating to contributions and support of environmental programs within the community, as well as on a national level. Among the initiatives that have been implemented by J&J under this phase is the creation of a management team team receives specialized training on how to communicate effectively with members of the community and how to present accurate, technical information in a clear and comprehensible manner. Phase IV also advocates facility involvement in community and professional organizations by encouraging facility personnel to invest their personal and professional expertise to these organizations as well as to the community.

Packaging Initiatives

J&J uses computer technology to project the impacts of changes in its packaging materials, to calculate instantly both the competitive and environmental impacts of design and material changes across entire product lines.

The system, called PackTrack, is a part of the company's Community Environmental Development function. PackTrack is a kind of "super spreadsheet" that allows the company to quantify and analyze the impacts of changes in its packaging materials, using the CONEG Preferred Packaging Guidelines as a model. For example, a user can input certain variables to test the impacts of a source reduction in a packaging material, and the spreadsheet program can generate reports by brands or product lines indicating annual cost savings, waste reduction in pounds per year, ratios of packaging to product materials, competitive comparisons, and other data useful in decision making. The system allows the company to pose "what if?" queries, such as, "What if we reduce heavy metals in package printing?" and understand the marketing and business impacts of the move.

Having developed PackTrack internally, J&J owns the copyright to Pack-Track. Interestingly, the company plans to donate the technology to other companies as part of its overall environmental commitment.

This and other uses of technology, such as the PRECIS emergency response system as described above, allow J&J to give quantifiable measures to its environmental policy while linking the company electronically to other companies and its surrounding communities. In this way, J&J is moving closer toward the concept of the industrial ecology of the future.

Environmental Commitment

J&J has learned from its Environmental Responsibility Program that it takes time, effort, and corporate commitment to get meaningful results. The company has learned that communication is the driving force for the entire program. An environmental policy understood by all employees and by the residents of the communities in which the company operates serves as the foundation for environmental leadership in the future. In the period the company has been developing and implementing this program, events throughout the world have demonstrated the soundness of its ideas and have affirmed its commitment to quality and responsible environmental management.

Changes are often complex, sometimes costly, where the direct benefits may not be obvious in terms of dollars and cents. However, J&J believes that it is essential to develop a long-term strategic plan and maintain a clear vision on where the company wants to be in the future, in order to achieve its goals and remain competitive in the "environmental 1990s."

11

A Venture in Environmentalism: The National Polystyrene Recycling Company

The emerging national consciousness of the municipal solid waste crisis in the late 1980s and early 1990s produced an unlikely focus of attention: polystyrene plastic. Comprising only 1 percent of the country's municipal solid waste in landfills by volume, polystyrene has been at the eye of a political storm that rose to a crescendo in 1990 with the decision by the McDonald's restaurant chain to drop the foam clamshell container.

Meanwhile, the polystyrene industry has been actively working to establish a recycling infrastructure. With funding by eight leading manufacturers of polystyrene plastics, the National Polystyrene Recycling Company (NPRC) was established in 1989 to create a program of responsible corporate citizenship. Through the NPRC, the industry made a commitment to demonstrating its dedication to manufacturing environmentally friendly products, helping collect those products after they are used, and recycling them into new products to be reused over and over again.

How well has NPRC met its goals? Does the McDonald's decision spell economic disaster for the polystyrene industry? And how did polystyrene become so controversial in the first place? The answers offer lessons for the entire packaging industry.

Formation of NPRC

At Dow Plastics, K. D. (Ken) Harman was business director for the company's polystyrene business. In the latter half of the 1980s it became increasingly apparent to Harman that polystyrene was going to be a lightning rod for environmental scrutiny. He saw that there would be nothing more important to his business than solving the environmental issues surrounding polystyrene.

A turning point came in a meeting of the Food Service Packaging Institute in 1987, a trade group representing disposable plastic and paper food packaging. There was a groundswell of interest in municipal solid waste management issues at that meeting. Harman seized the initiative by pledging $5,000, on behalf of Dow Plastics, to set up an environmental action committee, and he challenged his colleagues to do the same. Seven other companies stepped forward. This committee was to be the basis of the National Polystyrene Recycling Company, which was formally established in the following year, with eight members.

In his role as Dow Chemical's polystyrene business director, Harman participated in a February 1989 meeting with McDonald's, in which that company's upper management issued a challenge to the plastics industry. Noting it had been coming increasingly under fire for its use of plastics, McDonald's enlisted the help of the plastics industry in responding to public pressures. It also set a time frame for another meeting six months later, at which McDonald's would review the progress they had made.

Harman saw this challenge as a personal mandate. He believed that polystyrene recycling was the answer, and he spearheaded an industrywide effort to build a recycling infrastructure. He contacted his seven counterparts among the polystyrene resin producers and proposed the establishment of a polystyrene recycling company, a concept that was refined into what became the NPRC.

Probably none of the eight founding members had anticipated the magnitude of the investment, both in dollars and human resources, that would be needed to make this recycling infrastructure a reality. By early 1991, each member company had contributed $4 million to the effort. Nor did anyone anticipate the travel time, the meetings, and the sheer "sweat equity" that was required. Indeed, the NPRC was moving into uncharted waters. What made this remarkable achievement possible was a shared vision and a clearly understood goal: Create an infrastructure to recycle 25 percent of the polystyrene produced for food service and packaging by 1995.

The NPRC is not a trade organization but a for-profit company, whose initial mandate was to establish recycling facilities across the country. The company had barely been launched when it was called upon to communicate its goals on national television. After one week of existence, Harman was representing the NPRC on the *Today* show. Acting as the spokesperson for eight competing companies was a daunting task, Harman recalls, but it is a role he has assumed many times since.

Harman is quick to credit Dow for giving him the latitude to help establish the NPRC, and he credits the other seven resin producers for their support. He is convinced that if Dow and the industry had not initiated this effort, the company's polystyrene business would have been negatively impacted.

With the October 1989 purchase of Plastics Again, a polystyrene recycling facility in Leominster, Massachusetts, the NPRC's recycling infrastructure had broken ground. A second facility was added in the Los Angeles area in 1990, and three more in Philadelphia, Chicago, and San Francisco were in the final stages of construction in 1991.

While the NPRC has a small and lean staff, its eight parent companies offer the technical capabilities to understand what needs to be done. However, the backing of eight corporate powerhouses does not make the NPRC's task any less challenging. The group has focused on one issue at a time. Its first step was to acquire or build recycling plants, to demonstrate that it was technically feasible to recycle polystyrene. Two years into this effort, the NPRC could only begin to look at sourcing issues—creating a steady supply of post-consumer polystyrene. By 1990, the NPRC had appointed eight sourcing managers whose sole function was to develop polystyrene sources, working with schools, businesses, and large institutions to educate them about recycling and to establish collection efforts.

A final important function—communications—was assigned to a public relations firm until an internal communications function was established in 1991. Both the internal function and the outside agency field inquiries and supply literature on polystyrene recycling. They also developed an issue-oriented advertising campaign on foam packaging.

Based in Chicago, the NPRC is a low-overhead venture, with most of its funding devoted to the establishment of recycling plants. NPRC has been faulted for its low investment in communications and outreach activities, but it must be remembered that NPRC's role is that of a manufacturing company, not an issue management group. Much of its marketing is a grass-roots activity and is part of the entrepreneurial nature of the company. In addition, issue management support is provided by a Washington-based organization, the Polystyrene Packaging Council, which is devoted exclusively to communications programs and issue management.

Education is emerging as perhaps the most serious challenge facing the NPRC, and indeed the entire polystyrene industry. Polystyrene foam, especially in its most visible application in the food service industry, has been under siege by environmentalists and has been the object of innumerable public misconceptions. Before analyzing some of these perceptions and objections, it would be worthwhile to review what polystyrene is and how it is made.

Polystyrene: Perceptions and Realities

Styrene monomer, a clear and colorless liquid composed of hydrogen and carbon atoms, was first produced synthetically by French scientist Pierre Berthelot in the nineteenth century, and it was first manufactured in quantity during World War II.

Polystyrene is made when the molecules of styrene monomer are linked together using heat and pressure to form longer chains of molecules called polymers. The resulting polystyrene plastic can be molded into a variety of consumer products, clear food packaging, plastic dinnerware, tape reels, TV casings, office supplies, medical ware, and many other uses.

Polystyrene foam is manufactured with a blowing agent to produce small white beads known as expandable polystyrene, or EPS. These beads are expanded and molded or extruded into the polystyrene foam commonly used in packaging materials and disposable drinking cups, as well as in building insulation.

Perhaps the most common complaint against polystyrene is over the supposed use of chlorofluorocarbons (CFCs) as blowing agents. In reality, most of the foam packaging in the food service industry, including coffee cups, was never produced using CFCs. In 1988, after the Environmental Protection Agency indicated that fully halogenated CFCs could be responsible for depleting the earth's ozone layer, all foam packaging producers committed resources to shift to alternative blowing agents, of which the most common are pentane and butane.

Some have argued that these alternative blowing agents are not much better than CFCs. Pentane and butane are classified as volatile organic compounds, or VOCs, which can produce smog when emitted into the atmosphere. To minimize this possibility, federal and local regulators require manufacturers of polystyrene foam to limit factory emissions of VOCs. In most manufacturing processes VOCs are captured and routed through the boiler system to be used as fuel.

The industrial pollution argument was one of the top concerns of the Environmental Defense Fund (EDF), the group that advised McDonald's to drop the foam clamshell. According to EDF's senior scientist, Richard A. Denison, polystyrene's primary feedstock, styrene, was one of the most prevalent toxic chemicals released into the environment in 1988. [1]

It is true that traces of styrene have been found in the fatty tissue of all age groups and all geographical regions of the United States. However, there is no evidence to link this with polystyrene foam products. Because styrene is used in the manufacture of polystyrene foam products, the U.S. Food and Drug Administration has placed a ceiling of 10,000 parts per million (ppm) of residual styrene in food service products. Actual levels in foam food packaging are typically less than 1,000 ppm, and the FDA has declared polystyrene foam safe for use in food packaging.

Recycling Issues

Some technical issues remain in polystyrene recycling. While the early problems with cleansing the material have been resolved, densification to improve transportation economics remains a challenge. The NPRC's approach is to learn by doing. "If we start today, we'll find a better way tomorrow. That's where we are in this program," Harman said.

A more important issue is material collection. Because very little polystyrene waste is generated by households, curbside collection is only marginally helpful. Nor do mandatory recycling laws tend to boost polystyrene recycling in a significant way. The NPRC is looking instead to large institutional food establishments, such as school and corporate cafeterias, as well as to large automotive and electronic plants that ship parts in polystyrene foam.

The NPRC's Los Angeles plant is a $5 million, 60,000-square-foot facility that handles 13 million pounds of polystyrene a year. The Los Angeles school system uses 2.5 million pounds of polystyrene annually, and the NPRC facility is collecting from every school in the district.

Another major issue is markets. The NPRC sold 3 million pounds of recycled polystyrene feedstock in 1990 for one application alone: plastic luncheon trays. Other major market areas that the NPRC is working to develop are videotape cassettes, toys, and other household items.

The NPRC defines one of its challenges as that of changing polystyrene recycling from a supply-driven business to a demand-driven activity. It is pushing for durable product manufacturers to become more involved in using recycled materials. At a time when polystyrene food packaging was coming under intense scrutiny, virtually no public attention was paid to the durable consumer goods made from polystyrene. Indeed, many consumers would be surprised to know that their television sets and other home appliances have cabinets that are probably made from polystyrene.

The McDonald's Decision

McDonald's had staunchly defended its use of polystyrene food packaging for its ability to keep foods warm, its sanitation and safety, and its low cost, which helped keep prices down. In addition, the company argued that polystyrene is environmentally sound because it is fully recyclable, and it made a public commitment to helping customers recycle its polystyrene clamshell containers. McDonald's began working with the NPRC on pilot projects in 1989 and had planned to announce in the fall of 1990 that it would take its recycling program national; 8,500 restaurants would provide polystyrene feedstock for the NPRC.

This strategy came under increasingly aggressive and sometimes violent attack. In October 1990, environmental protesters smashed the windows of a San Francisco McDonald's restaurant because of its use of polystyrene. The Pro-Environment Packaging Council, formed with the backing of major paper companies, enlisted the aid of women's clubs and teachers to tell schoolchildren around the United States about the dangers of polystyrene and to encourage school letter-writing campaigns. Another group, the National Toxics Campaign, encouraged schoolchildren to mail their empty clamshells to McDonald's with angry letters. At one point, McDonald's was receiving 1,000 protest letters a day from schoolchildren.

McDonald's and the Environmental Defense Fund formed a task force in August 1990 to study municipal solid waste issues. It was a partnership that surprised industry and environmentalists, and both parties were criticized for "consorting with the enemy." However, the real surprise came on November 1, 1990, when McDonald's announced that it would begin a complete phaseout of polystyrene foam packaging as the first step in an environmental initiative.

Environmentalists hailed the action. However, some industry observers suggested that McDonald's had caved in, succumbing to a form of environmental blackmail. EDF has argued that the alternative packaging, paper-based wraps, represents a 70 percent reduction in landfill volume relative to the foam clamshell.

The McDonald's decision did not affect the NPRC's determination to push forward with its goal of recycling 25 percent of the polystyrene used in food service and packaging applications annually. NPRC's greatest loss was the education value of a highly visible restaurant chain's supporting recycling. This was far more important than the value of the polystyrene feedstock McDonald's would have provided, which, in reality, was rather low in volume. (A typical McDonald's restaurant produces a mere 20 pounds of polystyrene waste per day, and half of that is lost to take-out customers. In addition, NPRC had found in its pilot projects that the McDonald's consumer is not always scrupulous about separating waste properly, and about 35 percent of the McDonald's waste returned as polystyrene feedstock was actually paper and food waste. These contaminants were expensive to sort at the recycling plant.)

NPRC's Harman admits that, given the choice, he would have preferred the education and public relations value of the McDonald's partnership in recycling. McDonald's decision in itself brought renewed public attention to polystyrene recycling, and an increase in requests for information from the NPRC. Even though polystyrene is a small contributor to the municipal solid waste problem, each part of it deserves attention, Harman says.

Lessons From the McDonald's Story

While the McDonald's decision was widely praised in the popular press, it was criticized by advocates of recycling. On one point most observers could agree: The decision was successful in stopping the deluge of letters from schoolchildren. As McDonald's president Edward Rensi told *Forbes* magazine, "Although some scientific studies indicate foam packaging is environmentally sound, our customers don't feel good about it." [2]

The NPRC's Harman confided that the polystyrene industry could have done more to support McDonald's use of recycling. "We could not get the polystyrene industry together to support a major education effort," he said. As other corporations have learned, building an infrastructure is not enough; it has to be matched with an equally formidable public awareness effort.

"If I were sitting in McDonald's shoes, I would have made exactly the same decision. They wanted the letters from school kids to stop. Did they agonize over their decision? They sure did. But they did it with full knowledge of the arguments on both sides," Harman noted.

The NPRC has no doubt that plastics will play a positive role in the environment because of recycling. However, it has no illusions about the investment of time and money that will be needed to make a national recycling infrastructure a reality. Nor does it have doubts about the hurdles involved in building an industry consensus, even when the goals are quite clear.

The NPRC has deservedly won praise for its initial efforts to build a recycling infrastructure. But the polystyrene story has yet to be played out. Perhaps the most important lesson from the NPRC is already clear: It is not enough to solve the technical issues while ignoring the human factor. As one industry executive put it, "Environmental issues are emotional. Environmental solutions are technical. But environmental decisions are political."

Notes

1. Richard A. Denison, "McDonald's PS Ban Fosters Good Faith," *Plastics News*, December 10, 1990, p. 9.
2. Phyllis Berman, "McDonald's Caves In," *Forbes*, February 4, 1991, p. 73.

Conclusions
and Strategy

Today, the notion of a partnership between industry, on the one hand, and environmentalists and the general public, on the other, is greeted with hesitancy and suspicion on all sides. There is a language barrier of enormous proportions, an all too frequent mismatch between the consumer's image of environmental concern and what industry considers as environmental concern.

Historically, many industries, from raw materials to mining to forestry to oil production, have paid a heavy price to get their environmental houses in order. All have dealt with environmental issues. A giant paper company, for example, may feel it has done its part for the environment by investing millions of dollars complying with clean air and water standards; it may have devoted years of research to forest management techniques, such that it can now reforest an area in half the time that nature can. This is not a hypothetical example; the United States has more acreage devoted to forest today than it did at the turn of the century. Yet many companies are reluctant to go public with environmental initiatives. In going public they run the risk of being targeted for things they are not doing, rather than getting credit for what they have accomplished.

The consumer and the environmentalist speak from a very different experience than industry's. From their perspective, industry is the polluter of lakes and streams, the destroyer of the ozone layer, the creator of dioxins, PCBs, oil spills, and the Love Canal disaster. Seldom recognized is the fact that the consumers' concerns for their families' health and safety, as well as the environmentalists' concerns for the protection of wildlife and nature, are shared by many business leaders, who may live in the same communities. If put into the same room together, the environmentalist and the industrialist might find they agree on many of the same goals. By taking a "friendly adversary" approach the two sides could enter into a dialogue to arrive at rational trade-offs between the socially acceptable and the economically necessary.

We have examined a number of strategies companies have adopted to minimize the impact of their packaging on the natural environment. Yet we cannot say with certainty what an environmentally friendly package is. We cannot point to environmental truths that are static and immutable, across all regions and all times. What may not be recyclable today may be easily recyclable tomorrow. Processes that voraciously consume energy today may be energy-efficient tomorrow.

Many of today's assumptions about packaging and municipal solid waste management will change over time. Landfills, once considered a benign way to deal with waste, may emerge as the environmental nightmare of the 1990s, as more and more aging landfills are identified as contaminating groundwater supplies. Assumptions about degradability—hardly a useful concept when waste is sealed in landfills—will change dramatically if composting becomes more viable and widespread. Waste itself may ultimately be a meaningless term, if all that is discarded is increasingly re-used, re-cycled, or converted into energy. Perhaps the one packaging principle that will stand the test of time is source reduction. Whether we are dealing with aseptic juice boxes or plastic milk jugs, reducing or eliminating waste at the source will be a concept of lasting value.

Meanwhile, as the rules of the game constantly change, the challenge will be to balance the packaging needs of the consumer—for food preservation, product protection, economy, and life-style considerations—against the best current thinking on environmental practices. Industry cannot meet this challenge alone. The best hope for progress lies in communication and consensus building among all the key stakeholders—consumers, government, retailers, educators, the media, environmental activists, and industry.

This is not to say that there will be agreement among these players. Indeed, the adversarial role of activists in a free society has brought environmental progress that has been unknown in closed societies. Friendly or not, conscious or not, a loose partnership exists among these seven constituencies. Each of us has a role to play in this complex system of checks and balances, and knowledge is the best resource we can bring to the table.

Here are some final thoughts on what each of the key stakeholders has to contribute toward environmental progress in packaging. And because we believe that industry is best positioned to provide leadership in these complex technical issues, we provide specific guidelines for business leaders.

Consumers

As we have seen, consumers in survey after survey report an interest in environmentally sound products and packaging, although their actual buying practices do not always match their intentions. Where environmentalism is concerned, consumers fall into four categories:

1. *The "deep greens."* Consumers who proactively seek out green products and are vocal in their demands for environmental safety. Many of these belong to environmental organizations.
2. *The "medium greens."* Those who may not speak out, but if presented with the issues, will usually choose the green product. They may recycle at home, while in their purchasing, cost and convenience may still override environmental concerns.
3. *The "light greens."* Consumers who do not factor the environment into their thinking or their purchase decisions, but may be motivated to do so if presented with the right choices and opportunities.
4. *The "anti-greens."* A minority of consumers who are hostile to the environmental movement and consider it an intrusion on their personal rights and freedom.

In addition, there are regional differences among consumers. An Arthur D. Little survey of U.S. consumers found that the greatest percentage of deep green consumers are located on the West Coast and in the Northeast—fewer in the Midwest and the fewest in the South. Purchase decisions in the South were based more on issues of life-style and convenience than on environmental issues. These consumer attitudes tend to correspond with the regions most urgently affected by environmental crises and landfill shortages, and will surely shift as the municipal solid waste crisis hits more and more consumers closer to home.

Once the environmental issues become personalized, they will become very powerful motivators. As the pharmaceutical industry discovered in the area of tamper-resistant containers, consumers will change their buying behavior when personally touched by issues of health and safety. This personalization tends to be deepest for those consumers who live in the region where a disaster has occurred. Local resources and climatic considerations can also be motivating factors, for example, in areas where water shortages drive up costs and force conservation.

Perhaps the greatest opportunity for change lies among the so-called light green consumers. The deep greens and anti-greens have made up their minds already, and the medium greens will pick up their cues from the deep greens. These three groups will take it upon themselves to become informed, to the extent that it suits them. However, the largest group by far is the light green group. They are somewhat complacent, well-intentioned, but poorly motivated.

This light green majority is rightly confused by conflicting information. Most consumers cannot be expected to know one polymer from another, or to become so immersed in life cycle assessment as to know whether paper or plastic containers consume more energy. This group needs to be motivated rather than educated—for education implies that it is feasible for the large majority of consumers to become expert on deeply technical issues. There needs to be a motivation to do the right thing and the infrastructure in place to act upon those motivations.

Cost is one clear motivating factor. If the reusable, refillable, or recyclable package costs less, it increases the motivation to buy it. If garbage collection costs more than recycling, consumers will turn to recycling. And cost has long been a factor in frugal housekeeping behavior that unintentionally benefits the environment.

Peer pressure is an additional motivating factor; when most residents in a neighborhood are filling their curbside recycling containers, the pressure is on others to follow suit. The same kind of pressure that successfully turned the public tide against cigarette smokers and "litterbugs" can also be brought to bear against the nonrecyclers and the throw-away society.

Industry can play an important motivational role in providing products with no packaging, minimal packaging, or packages that are easy to reuse, refill, or recycle. Public service announcements can provide additional encouragement to move consumers from the light green to darker green categories. However, no inducements will succeed if the "green" package does not perform as well as the package that makes no such claims. Companies will do a disservice to the environmental movement unless their green initiatives offer equal or improved value to the consumer.

The Media

The technical, trade, business, and popular media represent a variety of viewpoints and readerships. The popular media in particular have been criticized for focusing on the negative, covering crises and disasters rather than positive environmental achievements. The least heard among all these is the objective journalist who can bring multiple viewpoints to bear.

We must remember that the media are an industry. They each have different products to sell to their markets, and like other industries, they will adjust their products to maximize sales. If a magazine or newspaper perceives that its audience craves the sensational, whether pro-environmental or pro-industrial, it will mold its product accordingly.

Educators

Educators face an enormously challenging task in researching and teaching the complex issues surrounding environmental responsibility. Much can be accomplished in the grade schools; Du Pont's grade school recycling program, which uses a green kangaroo mascot called Recyclaroo, provides one example of how to make complex issues accessible even to the very young.

Environmental issues are appropriately finding their way into curricula at every level, from preschool through college and postgraduate levels. The grade school years are an excellent time to instill environmental values, and school programs are being reinforced by excellent children's television programming. Community programs sponsored by such groups as the Girl Scouts, Boy Scouts, and extracurricular environmental clubs provide excellent support to science class teachings.

And environmental thinking is adding new courses and revising the parameters of existing courses at the college level, from business schools to engineering to industrial design. Ideally, environmental issues will not be treated as separate or distinct, but will be integrated into the basic premises of scientific and economic thought. In this way, environmentalism will be a lasting component of higher education, and not a short-lived fad.

Government

Remarkably little environmental legislation is initiated at the federal level. Most of the major federal environmental laws are a modification of state laws. Smart companies are thus paying more attention to state and local legislation efforts. As we have seen, Procter & Gamble provides one example of a company that has worked effectively with local authorities on municipal infrastructure issues, and hundreds of other companies are working with local authorities to help shape laws that directly affect them. In fact, by the time a new bill is debated at the federal level, the battle at the local level may have already been decided.

As the regulators do their job of enforcing legislation, translating law into action, it is often only then that the unintended costs and side effects of a law are revealed. Regulators then pressure legislators, in turn, to rethink their policies. In this sense, regulators are more of a driving force than legislators.

Corporations have long complained that regulators give mixed signals. It is impossible to predict the direction of future legislation, and companies cannot factor future regulations into their long-range plans. As we saw in the example of 3M, public policy leaders took a stand on pollution prevention in the mid-1970s—and then shifted course in favor of pollution cleanup. Surely, providing incentives for pollution prevention is a wiser course than allowing environmental damage, then applying punitive measures after the fact.

The EPA municipal solid waste management hierarchy that makes source reduction the top priority is an excellent model, and one that will undoubtedly pass the test of time. Source reduction parallels pollution prevention; the basic concept is to prevent a problem before it happens. Unfortunately, too many legislative initiatives have focused on recycling at the expense of source reduction. Because the aseptic juice box is a challenge to recycle, it has faced bans in several states; the source reduction value of the package has often been overlooked.

For too long, industry has viewed any investment in working with legislators and regulators in support of environmental issues as a non-capital-producing expense. Many companies are now challenging that view, and realize the economic advantages of partnership with the public sector. The Coalition of Northeastern Governors (CONEG) has provided an important model for industry and public-sector cooperation.

Environmentalists

Like the citizens and members they represent, environmental groups vary widely in both their goals and tactics. Some are chartered to preserve and protect the world we live in; others are advocates for change. Some merely challenge corporations to solve environmental problems; a rare few go further and also want to be part of the solution.

More often than not, business leaders and environmental activists have the same goals. Both want a clean, safe world for their children. From the environmentalist's perspective, business moves too slowly toward achieving this cleaner world. From the business point of view, environmentalists seem to make unrealistic demands.

There are no perfect solutions, just as there is no such thing as a risk-free society. Business leaders have to work within an industry structure and deal with many forces. They are responsible to shareholders who have invested in their companies, and to workers who depend on them for jobs. A paper

company can make a massive investment in cleaning the wastewater effluents from a pulp-bleaching process and still face ecologists' demands for a
total shutdown of the plant. There has to be a middle ground.

Partnerships between industry and environmental groups may not always
be feasible; many activist groups have charters and funding requirements
that prohibit them from "consorting" with industry. And single-issue groups
are especially ill-suited for partnering with industry to solve complex issues.
For example, the growing "back-to-nature" movement in Europe, which
shuns technology and all the trappings of a consumer society, would find
little common ground with most large consumer product companies.

For their part, environmentalists walk a fine line. They cannot be perceived as endorsing a product or a company, and this creates a hesitancy to
praise or support a corporation that is achieving environmental improvements. Surely, though, the time has come to establish a common ground
between industry and environmentalists.

Retailers

A few national retail chains have been cited again and again as examples of
an environmental trend in retailing. The best known are WalMart in the
United States, Loblaws in Canada,and The Body Shop, based in the United
Kingdom. We should point out that there are thousands of retail chains in the
United States alone and that these few examples are only early indicators of
an emerging environmental trend in retailing. Each of these retailers was
largely driven by social consciousness, and each required an executive-level
decision to develop guidelines for suppliers and environmental product
sections; however, it remains to be seen if these initiatives will continue if
profits do not justify them.

Green marketing can be used competitively as a differentiating feature,
especially on a regional or local basis, to cater to pockets of strong environmental sentiment. The Massachusetts-based Bread & Circus, described in
Chapter 1, is one example of a regional health food chain providing leadership on environmental issues.

However, for most retailers, any product, green or not, faces stringent criteria to justify shelf space. The principles of DPP—direct product profitability—remain the driving force of retailing.

In the end, the manufacturer has greater influence over environmental progress than most retailers. And much of the staying power of green products will ultimately depend on the buying practice of the consumer.

Industry

We will consider two sectors on the industry side—the waste management business and the packaging companies.

More people are looking to waste management companies as a key player in environmental progress. Through alliances with industry, such as the Du Pont and Waste Management, Inc., joint venture to recycle plastics, waste management companies are helping to influence the process, to establish markets for recycling, and to create an infrastructure.

The waste management companies can influence initiatives at the local level, through what they choose to collect, as well as at the national level, through major investments in incineration, collection, sorting, and processing technologies. They can influence what is recycled and even what materials are used. For example, German chemical companies face legislative pressures to take back packaging used for chemicals. An entrepreneurial waste management company might offer to collect those containers for the chemical company, and thereby play a determining role as to what types of containers and materials are used. It can act as a force in market development, as well as a facilitator for government efforts to put incentives and disincentives in place. Waste management companies have an opportunity to develop a proprietary position in local or national infrastructures.

The packaging industry itself is where we see the greatest opportunity of all to take a leadership role. Because of the enormous complexity and diversity of the packaging industry, no other player is better versed in the technical, economic, and performance issues of packaging. No one is better positioned to influence every facet of the supply and demand chain, through its purchasing decisions and marketing dollars. And as we have seen in Chapter 2, many packers have already pursued environmentally sound policies for years, from recycling to lightweighting to source reduction to multi-functional, multiuse packaging.

Opportunities for Partnership

How will it be possible to bring these diverse constituencies into a partnership for progress?

There needs to be a partnership mentality so that responsible initiatives may be put forth and allowed to be nurtured—and not killed by protest before the desired results are achieved. The environmentalist, by engaging in a dialogue with the businessperson who has done a responsible job, can productively bring pressure to bear to achieve results.

We need to establish a common language that brings together environmentalists and industrialists so that they can achieve consensus as they address environmental issues from their different perspectives. Environmentalists did not originate as people trying to solve industry's problems, and industry did not originate as a force for preserving the environment, but a new breed has emerged. People who can talk the language of both sides and bridge policy issues in a way that makes sense will be much in demand in the future. The CONEG model, despite its ups and downs, shows what can be accomplished when industry, legislators, and environmentalists come together to develop environmentally progressive public policy.

Established consensus-building techniques do exist. For example, the National Coal Policy Board was formed in 1977 by two unlikely partners: Dow Chemical and the Sierra Club. Dow's former corporate energy manager, Gerald Decker, approached Larry Moss, former president of the Sierra Club, to resolve a long and costly dispute over the siting of a new power plant. Using a technique called the Rule of Reason invented by Milton Wessel, the two sides, which had been traditional adversaries, came up with more than 200 points of agreement—over an issue that had already cost both sides many years and millions of dollars in litigation. For example, both sides agreed the EPA should grant Dow a limited number of exceptions to the Clean Air Act to promote and encourage new technologies.

As we have seen in case study examples, there are companies throughout the packaging industry that have found ways to make environmental investments as sound business practice. The environmentalist should be building the same case. Many ecologists are technical economists as well as scientists of the first order. Their input—indeed, their opposing points of view—are vital to the process. A good example of what happens without constructive debate may be seen in the Eastern Bloc countries. The black squalor of Eastern Europe's smokestack regions is largely a product of state-controlled industry that had been unchallenged by political activism. There was no public debate, no state intervention, and no industry initiative to achieve clean air and water. It is perhaps the starkest illustration of the importance of dialogue between industry and the public.

Packaging is the single largest component of the municipal solid waste stream. But we cannot do without packaging. It is a vital contributor to the economy and to the quality of life. As we have noted, Third World countries lose between 30 and 50 percent of their food to spoilage and damage—for lack of suitable packaging and distribution systems.

Environmentalists and industrialists together need to identify economically viable, environmentally sound initiatives with long-term, systemic benefits, so that these may be factored into major capital investments in a complex system. In this way, the value of packaging to the economy and to society can continue, with the gentlest possible impact on the natural environment.

Industrial Ecology

In Chapter 6 we described the concept of "industrial ecology" as the only logical philosophical framework for the green corporation of the future. In sum, this concept is a recognition that the traditional industrial model—in which individual manufacturing processes take in raw materials and produce valuable goods plus valueless waste—should be transformed into a more integrated model in which the consumption of energy and raw materials is optimized, waste is minimized, and the effluents of one process are the raw material for another process.

Mining the value in waste materials can yield new business opportunities. Consider the following examples:

- Dow Chemical's WRAP (Waste Reduction Always Pays) program, formalized in 1986, says that its waste reduction programs pay for themselves within ten months. One program involves converting waste streams into specialty chemicals. In 1990, Dow reported revenues of $70 million from the sale of products derived from its major polymer intermediates; disposal of the same products would have cost the company about $25 million.

- Du Pont ceased ocean dumping of acid iron salts, in response to public protests, even though its scientists considered the practice harmless; it found a new market for the salts in water treatment plants. In addition, the company is marketing the expertise it has gained in environmental issues by helping its customers clean up toxic wastes, a new business for which it forecasts revenues of $1 billion by the year 2000.

Our discussions of materials and infrastructures have illustrated some of the issues and obstacles in creating a sustainable industrial ecosystem. While an integrated waste management system—embracing recycling, waste-to-energy conversion, composting, and other options—is technologically feasible, there must be an infrastructure in place to accommodate these options and ready markets for the usable materials. However, these obstacles only further demonstrate the fact that tomorrow's "green" corporation cannot operate in a vacuum. There needs to be many synergistic industrial ecosystems with many interbusiness linkages, in which each process and product contributes to the effective functioning of the whole.

As Robert Frosch and Nicholas Gallopoulos, two senior General Motors research scientists, point out,

> *An ideal industrial ecosystem may never be attained in practice, but both manufacturers and consumers must change their habits to approach it more closely if the industrialized world is to maintain its standard of living—and the industrializing nations are to raise theirs to a similar level—without adversely affecting the environment.* [1]

Meanwhile, as responsible corporations look ahead to the industrial ecosystem of the future, there are steps they can take along the way to move us closer to the ideal. The following are some guidelines and conclusions drawn from Arthur D. Little's work on packaging issues in industry and the public sector, as well as from the collective research for the writing of this book. They are not intended as the definitive answers, but as achievable measures and realistic models that can be adopted in a rapidly changing world.

Guidelines for Environmental Strategies

Leadership

There are several models of business leadership for environmental progress; there is no "right" model, except that which is most appropriate to the culture of the individual organization.

Leadership may come from the top down; that is, top management defines the environmental missions and enforces and reinforces environmental policy. This enforcement is effectively accomplished by making environmental performance a criterion of performance reviews and compensation. A model of top-down leadership is Du Pont, whose CEO declares himself the company's "chief environmentalist." Top management needs to provide clear directions and goals. It can draw upon the profit motive, at least in the beginning, so that managers trained for decades in the profit motive can easily assimilate the environmental dimension in their operational thinking.

Leadership may also be grass-roots in origin, springing from anywhere in the organization, and this kind of leadership should also be encouraged. It may take shape in the form of a task force or committee, which may then deliver its thinking to top management—the Procter & Gamble model. Or, as in the case of 3M, it may fit within a culture that encourages teamwork and idea-sharing, often through nonfinancial incentives. However, no matter who first identifies the environmental agenda, it must ultimately have the full support of the CEO.

Whatever form it takes, corporate leadership must be committed. And, as with the consumer, the greatest commitment will come from those who have personalized the problem. For example, Procter & Gamble's top management involvement in local Cincinnati infrastructure issues sensitized them to broader environmental issues. This personalization of the issues may also be inspired by moral or ethical values, or it may be a reaction to an environmental disaster or to punitive regulatory measures.

Empowerment

To promote implementation of a sound environmental agenda within an organization, the most effective and least costly approach may be to incorporate environmental thinking into current programs, rather than to create new programs from scratch. Such proven and existing programs may include established cost reduction programs and total quality management (TQM) programs.

The TQM concept is emerging in a number of organizations as an ideal vehicle for environmental quality programs. TQM encourages the emergence of ideas in the organization from the bottom up, and it provides a logical framework for environmental objectives.

For corporate initiatives to have credibility, managers need to commit to them early and communicate objectives broadly. This process can make use of existing communication channels—in-house newsletters, slide shows, seminars, databases—and new information resources may need to be created.

There also needs to be an appropriate corporate culture or mindset in which long-term policy objectives are meaningful to the individual employee. Companies can create an attitude for long-term thinking with short-term action. Simple initiatives, such as recycling office wastepaper, help foster a mind-set with long-term ramifications, and one that empowers employees at all levels to participate.

Competitive Positioning

Creativity in product development and market positioning must reflect the environmental infrastructure available to the consumer. For example, Procter & Gamble, as we have seen, offers a variety of container concepts, allowing consumers to choose the ones that fit the waste management options available to them.

Companies may look for competitive advantage in packaging source reduction, recycling, and other post-consumer waste management strategies. They should also take full advantage of the sale or reuse of process waste—a common practice in supply-side industries such as plastic converters and paper companies, but an often overlooked environmental plus in demand-side industries.

The opportunity to develop product leadership through investment in the infrastructure or in unique sources of supply should not be overlooked. This investment may be in the form of strategic alliances, such as the Du Pont and Waste Management alliance to recycle plastics. It may also take the form of purchasing patents or licensing rights, such as Avery Dennison's acquiring the licensing rights for a polyethylene label technology, called Fleximage, which is fully recyclable with polyethylene packaging.

Companies should be sure to factor environmental positioning into capital expenditure formulas. Many environmental efforts will bring immediate cost advantages through efficiencies of operations, the reuse and marketing of process waste, and packaging source reduction. Once these agendas are accomplished, there will also be a need to consider long-term investments in which financial return may not be clear. Industry leaders are approaching environmental issues on faith, recognizing the value of corporate reputation. For other companies, such a leap of faith is a "hard sell" and out of sync with accepted ROI criteria. However, as we have seen in our case studies, it is faith—a belief in "doing the right thing"—and not financial criteria that has been the driving force for some of the most creative environmental positioning.

Finally, as noted earlier, competitive positioning cannot overlook the human factor. As one executive summed it up, "Environmental issues are emotional. Environmental solutions are technical. But environmental decisions are political."

Partnership

Environmental concern can often best be demonstrated to company personnel by taking leadership in local community issues. Approaching environmental programs on a grand scale—trying to "save the world"—may ring hollow to people on the line or the factory floor if, indeed, the company has made no commitment to the community in which it operates.

Companies should go beyond mere compliance with regulations and look to the local community's concerns and desires. 3M has enhanced its corporate reputation by going far beyond what the law requires. And Du Pont offers a valid reminder that the company's facilities exist "at the pleasure of the local community."

Companies can share their technical expertise in composting, incineration, and other waste management options with the community. They can help the community in siting issues—but the partnership should not stop there. Companies need to make a long-term commitment to follow-on participation, to ensure clean, effective operations.

As each environmental issue is identified and targeted, "partnering" should occur at levels appropriate to the problem. In other words, a community problem should involve community leaders, other local industry, and municipal organizations. A state or regional problem should involve state or regional leaders with regional or local representation, leveraging both the private and public sector of the region. National or global programs should be reserved for national or global problems.

There are also enormous advantages in industry partnering with educators. Support for educators needs to be thought through as carefully as any strategic corporate positioning dealing with consumers. The content of the education must be valid; it must first of all serve the learner and not be self-serving in terms of corporate goals. It must be appropriate to the level at which it is targeted: symbols, slogans, and teaching materials for grades K–6; apparatus, programs, and curriculum materials for grades 7–12; and research grants and other support mechanisms to encourage sound programs at the university level.

Just as there needs to be a partnership among the key stakeholders to resolve environmental issues there must also be a partnership mentality within the corporate organization. Environmental programs cannot be sustained if they are operated in isolation. A parallel need for partnership to make R&D organizations effective, was identified in the recent book *Third-Generation R&D*. [2] And 3M demonstrates environmental leadership achieved through the partnership of process engineers and environmental management. Such a partnership will call upon the resources and imagination of the company's sourcing (purchasing) organization, product design, R&D, manufacturing planning, processing/manufacturing operations, logistics planning, marketing, regulatory affairs, and finance.

Finally, corporate environmental philosophy and accomplishments have to be communicated—to consumers, legislators, the media, educators, and all the key players. Companies need to leverage the positive in making their environmental achievements known, but they should stop short of overselling, since exaggerated "green" claims may backfire. As one executive put it, "If you have a good story, tell it. If you don't have a good story, get one."

Linkages

Employees who become too focused on producing a product may see anything outside the realm of their immediate task as disruptive. The traditional functions within the corporate organization should be called upon to look further upstream in their supply chain and further downstream in the use of their product and its components. This approach can expand considerably the dimensions traditionally encompassed by a manufacturing planning, production, or other such function within a company.

Companies can create markets for recycled materials through their own purchasing dollars. Doing so requires a realistic understanding of the product's minimum requirements for packaging materials. For example, the company should evaluate recycled or recyclable materials in the light of essential packaging criteria such as preservation, product safety, and presentation. Packaging quality and performance must not be overlooked; environmental progress will be ill-served by shoddy products.

Having established an opportunity to use recycled or recyclable materials, corporations should then provide clear and concise specifications, communicating these basic requirements to suppliers. In this way, they can encourage innovation in the use of materials, assist suppliers in facilitating their needs, and reward the suppliers' efforts through their purchasing dollars.

The value of environmental positioning should be reflected in product development and market strategies. However, environmental technologies can and should be shared. And companies should challenge others in their industry to pursue environmental initiatives.

Closing Thoughts

In striving to achieve environmental responsibility, society faces daunting tasks. New technologies and infrastructures, as well as fresh insights are needed. Government must create adequate incentives for industry and take advantage of industry's know-how. Corporations must play an untraditional and sometimes uncomfortable role, acting cooperatively with other corporations. Even more challenging will be the international cooperation needed to address environmental issues that, by nature, transcend national borders. And all the stakeholders in the process need to abandon strictly adversarial postures and recognize common problems.

In spite of the challenges, there is no alternative. Industry can either practice a sound environmental strategy, and reap competitive advantage in the process, or be forced into playing a reactive or follower role, at a far greater cost. Indeed, the costs of inaction can no longer be seen from the individual corporate perspective; the costs and consequences of our actions have taken on a global dimension.

Among the partners, the packaging industry is best-positioned and equipped to take the leadership role and effect meaningful environmental progress—by virtue of its investment in technology, its deployment of disciplines, existing establishments and infrastructures, and the breadth of its influence all along the supply and demand chain. All of us, as stakeholders and partners, now need to support the packaging industry as it leads the emerging partnership for progress.

Notes

1. Robert Frosch and Nicholas Gallopoulos, "Strategies for Manufacturing," *Scientific American*, September 1989, p. 144.
2. Philip Roussel, Kamal Saad, and Tamara Erickson, *Third-Generation R&D* (Cambridge, Mass.: Harvard Business School Press, 1991).

Index

About the Authors

E. Joseph Stilwell
Joe Stilwell is Director of Packaging, Product Technology Laboratories, at Arthur D. Little, Inc. He has over 30 years' of diverse experience in the packaging industry beginning with his degree in packaging from Michigan State University. He has held management and consulting positions with Abbott Laboratories, Battelle, and Westinghouse. Mr. Stilwell holds several patents, is a frequent lecturer with academic and professional groups, and has served as packaging consultant to UNIDO. He has been instrumental in focusing Arthur D. Little's environmental technology and consulting resources on packaging issues.

R. Claire Canty
Claire Canty is a Senior Consultant in the North American Management Consulting Directorate of Arthur D. Little, Inc. She has more than 30 years' experience in the pulp and paper industry with her principal interest in the area of marketing and business strategy. Mrs. Canty's work has included studies of the economics, operations, and market prospects of paper and board mills, including the impact of municipal solid waste and recycling infrastructure issues.

Peter W. Kopf
Peter Kopf is a Director in the Technology and Product Development Directorate at Arthur D. Little, Inc. He has more than 21 years' experience in industrial research and development experience in polymer synthesis, characterization, and applications development, in addition to experience in specialty chemicals and solvents applications. Dr. Kopf has over 20 publications in the field of polymers and composites. He received his Ph.D. degree in physical chemistry at the University of Rochester in 1970.

Anthony M. Montrone
Anthony Montrone is a Director in the Environmental, Health, and Safety Consulting Practice and Manager of the Environmental Business and Strategy Group at Arthur D. Little, Inc. His work has included directing major life cycle assessment studies for leading corporations. Prior to Arthur D. Little, Mr. Montrone spent 12 years with the U.S. Environmental Protection Agency, and has extensive experience developing regulatory programs.

Sources for quotes at the front of the book

George Bush: quoted in information for The President's Environment and Conservation Challenge Awards (United States Government)

Gro Harlem Bruntland: writing in *Our Common Future* (Oxford: Oxford University Press, 1987)

Edgar S. Woolard: remarks made at the World Resources Institute, Washington, D.C., December 12, 1989.

James G. Speth: quoted by Robert L. Olson in *The Futurist*, May–June 1991

Paul E. Gray: writing in *Technology and Environment* (Washington: National Academy Press, 1989)

Heterick Memorial Library
Ohio Northern University

DUE	RETURNED	DUE	RETURNED
1. MAY 28	MAY 2 5 1993	13.	
2. FEB 1 5 '94	FEB 1 5 '94	14.	
3.		15.	
4.		16.	
5.		17.	
6.		18.	
7.		19.	
8.		20.	
9.		21.	
10.		22.	
11.		23.	
12.		24.